DATE DUE

THE BODY SILENT

THE BODY SILENT

ROBERT F. MURPHY

Henry Holt and Company   New York

Copyright © 1987 by Robert F. Murphy
All rights reserved, including the right to reproduce this
book or portions thereof in any form.
Published by Henry Holt and Company, Inc.,
521 Fifth Avenue, New York, New York 10175.
Distributed in Canada by Fitzhenry & Whiteside Limited,
195 Allstate Parkway, Markham, Ontario L3R 4T8.

Grateful acknowledgment is made to Grove Press, Inc.,
to reprint quotations from *Murphy*, by Samuel Beckett.
© 1957 by Grove Press, Inc. All rights reserved.

Library of Congress Cataloging in Publication Data
Murphy, Robert Francis, 1924-
The body silent.
Bibliography: p.
Includes index.
1. Murphy, Robert Francis, 1924–   —Health.
2. Spinal cord—Tumors—Patients—United States—
Biography.   3. Quadriplegics—United States—
Biography.   I. Title.
RC280.S7M87   1987   362.4'3   86-18455
ISBN 0-8050-0130-1

First Edition

Design by Kate Nichols
Printed in the United States of America
1 3 5 7 9 10 8 6 4 2

ISBN 0-8050-0130-1

*This book is dedicated
to all those who cannot walk—
and instead try to fly.*

CONTENTS

CONTENTS

III ON LIVING

PREFACE

This book was conceived in the realization that my long illness with a disease of the spinal cord has been a kind of extended anthropological field trip, for through it I have sojourned in a social world no less strange to me at first than those of the Amazon forests. And since it is the duty of all anthropologists to report on their travels, whether to Earth's antipodes or to equally remote recesses of human experience, this is my accounting. It has been written with many purposes in mind, but the most important is to relate to the general public, and to disabled people everywhere, the social circumstances of the physically impaired and the meaning of this condition as an allegory of all life in society.

No book of this kind has ever been written by an anthropologist, and it draws its primary inspiration from only one anthropological work, *Tristes Tropiques*—Claude Lévi-Strauss's towering narrative of his travels and research in the backlands of Brazil. The French anthropologist used his journey across geographic space as a backdrop and source for an inquiry into the structure of human thought. In much the same way, I have used my own odyssey in inner space to explore the structure of selfhood and sentiment.

The book draws on many sources. It is essentially a narrative of my own experience of paralysis, but my general understanding of this affliction has been enriched by readings on the social consequences of disability and by field research among paralytics. I was supported in this effort by grants from the National Institute of Neurological and Communicative Disorders and Stroke (NINCDS) and the National Institute of Mental Health (NIMH). Given the personal nature of this book and the fact that it does not make extensive use of project data, however, it cannot be considered, in a strict sense, as a report of the research program's results, which are being published in the usual scholarly formats and outlets. I am nonetheless indebted to the Institutes, although they are in no way associated with or responsible for the findings and views expressed by me.

Many people have contributed to the writing of the book. I have enjoyed the unstinting support of my colleagues in the Department of Anthropology at Columbia University, and the University has generously provided me with research assistants for the past ten years. The assistants, in order of their service, have been Brian Ferguson, Joel Wallman, and Steven Rubenstein; their friendship and support have made my continued professional life possible. My fellow researchers in the NIH-supported program—Jessica Scheer, Richard Mack, and Yolanda Murphy—were instrumental in developing my thoughts on disability during our three-year colloquy on the subject, and many of the ideas presented in this book were collective products. Dr. Stanley Myers extended the cooperation of Columbia-Presbyterian Medical Center's Spinal Cord Outpatient Clinic to us; he also contributed to both my education and my survival as one of my attending physicians. He was joined in this enterprise by Drs. Daniel Sciarra and Edward Schlesinger of Columbia-Presbyterian's Neurological Institute, and by Drs. Peter Desanctis, Francis Symonds, and the late Bard Cosman. They didn't succeed in curing me, for I am as fixable as Humpty-Dumpty, but they did help to keep me alive for the last ten years. As my book relates, I was for a while doubtful about whether or not this was for my good, but I now acknowledge it, with gratitude, as a gift.

Several people have read the manuscript of this book and have made valuable comments. Teri McLuhan has been a constant source of encouragement and support from the beginnings of the first draft through publication. Even the title was her invention. Other contributors have been: Louise Duval, Raymond McDermott, Katherine Newman, Katherine Pelletier, Barbara Price, Jessica Scheer, David Schneider, and Joel Wallman. And Jack Macrae, my editor at Henry Holt and Company, has done all the things that go with his territory and more, for he helped me carve a book out of the disorder and profusion of its early stages.

Finally, my wife Yolanda has been my constant companion and collaborator during this ten-year field trip, as she has been in all my other travels. She, more than any doctors or hospitals, has sustained my life—and enabled me to transcend simple physical survival and once again find purpose and joy in living.

Leonia, New Jersey   Robert F. Murphy
August 1986

THE BODY SILENT

PROLOGUE: NIGHT SOUNDS

*History is a nightmare from which I am
trying to awake.*

—James Joyce,
Ulysses

As the last of the evening's visitors trail off to the elevators, the neurological floor's ties to the outside are severed, and its inhabitants return to the closed world of the hospital. The floor differs from the rest of the institution, for its patients stay longer and many are deeply, irreversibly ill. They are not transients, here for one or two weeks, but habitués, denizens. Their confines are not alien to them—which makes the situation no less unpleasant—for the rhythms of their care are part of a familiar routine. Long-term patients become somewhat estranged from their other selves, and the people and events represented by their visitors lack for them the immediacy of what is happening on the floor. It is another dimension, and the reinstatement of the hospital regimen at nighttime allows the patients to fall back into well-worn paths, grown familiar through many hospitalizations. In a way, they are glad when their guests leave.

1

The floor stays busy as the aides distribute fruit juices and the nurses administer their ten o'clock medications, but by eleven the tempo slows. The inmates settle down to sleep, televisions are turned off, the hall lights are dimmed, and soon the only sound to be heard is the steady sufflation of a respirator or the rhythmic patter of a pulmonary therapist's hands as she pounds gently against the back of a patient with pneumonia. Occasionally, a phone rings at the nurses' station, but their voices become more muted as the night wears on. These are the quiet hours.

Night had settled noise into murmur one midnight when suddenly the calm was shattered by a screechy voice reciting the Apostles' Creed:

> *I believe in God, the Father Almighty,*
> *Creator of heaven and earth,*
> *And in Jesus Christ, His only Son, our Lord,*
> *Who was born of the Virgin Mary. . . .*

Silently I followed the prayer, a prayer I had not said since childhood, remembered not by words or meaning but rather by the cadences of its recital. Pounded into my head by nuns, I doubt that I ever reflected on its content, nor did I ever say it to myself. It was intoned in church and during catechism classes, chanted in the same rote singsong that still lodged in my unconscious mind along with all the rest of the detritus of my religious education. The prayer cast a momentary spell in its brief and atavistic resort to old forms, a spurious kind of solace that faded with the sounds.

My surprise at the prayer and my own reaction to it was not nearly as great as my astonishment at the source, for there could be no mistaking that it came from Katie, an Irish woman in the final stages of multiple sclerosis. Katie's body was so wasted that she looked like a featherless, wounded bird. On most days she sat in the solarium, lost in a big wooden wheelchair, her limbs splayed into a permanent rictus, uttering only unintelligible sounds. The sclerotic processes had robbed her body of both

2

volition and autonomic function and had ultimately deprived her of reason. I had no idea that she was still capable of coherent speech until the night of her prayer. I had thought many times about her and how much better death would be than to dribble out one's life in such a miserable condition. So, too, did Katie, for one night a few weeks later, her shrill voice called out, "Dear Jesus, take me with you and end my suffering."

The floor fell silent, the nurses and aides paused from their work, and I stopped breathing for an instant. Her prayer has since been answered, and death came as Providence, not Nemesis. But I knew from that moment that it would come as deliverance to me as well . . . in time.

This is the social history of a paralytic illness that has taken me slowly and inexorably from its first symptom, a little muscle spasm in 1972, to quadriplegia in 1986, the year in which these words are being written. The paralysis* results from a tumor inside my spinal column that is growing slowly but steadily and will eventually reduce my body to total quiescence. There are no medical miracles in this story, nor do I expect any in the future. But it has been an education of sorts, for in my passage into paralysis I have discovered the ebullience and power of the rage to live, which is really what this book is all about.

I started to write this account each year for the past four, and each time the project foundered on an inability to look upon myself as both subject and object of my observations, to act simultaneously as author and chief protagonist, to be both ethnographer and informant. There is, moreover, little reason in

---

* Many people object to the use of the terms *paralytic* and *paralysis*, for they are short and brutally descriptive, as well as redolent of the no-longer-used label of "infantile paralysis." I do this only for reasons of economy, in order to use one heading to cover a number of impairments: paraplegia (paralysis of the lower body and limbs); quadriplegia (paralysis of the lower and upper body, including varying degrees of atrophy of the arms and hands); hemiplegia (paralysis affecting one side of the body, most commonly due to stroke); and a variety of others not easily included in any of these categories.

my circumstance or career to warrant biography. I am a sixty-two-year-old professor of anthropology at Columbia University, the author of several books, and a teacher of graduate and undergraduate students for more than sixty semesters. I have been married for thirty-six years to the same woman, and we have two grown children. This is all strikingly unremarkable, and for this reason I will not dwell on my life and career, except when it is relevant. This is, after all, not my autobiography, but the history of the impact of a quite remarkable illness upon my status as a member of society, for it has visited upon me a disease of social relations no less real than the paralysis of the body.

To a certain extent, the social history of my impairment has much in common with that of other motor-disabled people, however much it may differ from them medically. Paralysis can result from spinal cord disease or trauma, brain damage, stroke, poliomyelitis, spina bifida, muscular dystrophy, multiple sclerosis, cerebral palsy, or any of a number of other conditions. And it can be congenital or can happen at any age, to people of widely different social circumstances. The similarity of our conditions is, however, social, for no matter who we are or how we got into our unenviable situation, the able-bodied treat the physically handicapped in much the same way. Disability is defined by society and given meaning by culture; it is a social malady. My account will, therefore, have considerable relevance for understanding the social lives of all disabled people. And to avoid the pitfalls of seeing the entire world in my own experience—the sin of solipsism—I will also draw upon my six years of reading and field research on the lives and times of the paralyzed.

The lessons to be learned from paralysis also have profound meaning for our understanding of human culture and the place of the individual within it. The relationship between society and its symbolic standards for acting and evaluating, on one hand, and the strivings and interests of ordinary people, on the other, are not neatly adjusted to each other and mutually supportive. Rather, the individual and culture are essentially in conflict, and history, instead of being the realization of human intentions and

4

cultural values, is commonly a contradiction of both. The study of paralysis is a splendid arena for viewing this struggle of the individual against society, for the disabled are not a breed apart but a metaphor for the human condition.

The disabled represent humanity reduced to its bare essentials, making them wonderful subjects for anthropological research. In 1937 the French anthropologist Claude Lévi-Strauss sought out the Nambikuara Indians of central Brazil, "looking for a society reduced to its simplest expression," but in the end found only human beings.[1] My own work reverses the direction of Lévi-Strauss's experience. I looked for human beings reduced by physical incapacity to a struggle for survival as mean as the Nambikuaras', and I found Society. Like Lévi-Strauss's Indians, the paralyzed are "marginal" people, and the study of their tenuous position at the edge of society will tell us much about all of social life.

In the long days and nights of my recovery from spinal cord surgery, I would lie in my hospital bed and think, too weak to concentrate on reading and yet not sick enough to watch television. And what I mused upon was myself and my new and permanently altered feelings of who and what I was. Gradually, my thoughts became disembodied, and I began to think of myself as if a part of me were perched over the headboard, watching the rest; it was as if it were happening to somebody else. Worry and foreboding were displaced in my reveries by absorption, and I became mesmerized by the magnitude of my disaster and a growing sense that it followed some deeper rhythm. And it is of this tempo, this structure, that I write.

Much has been said in recent years about the right to die, and specifically about whether death is preferable to severe disablement. A play, later made into a motion picture, was titled with the rhetorical question, *Whose Life Is It, Anyway?* The drama concluded that the hero, a young musician rendered quadriplegic by an accident, had a right to be allowed to die without medical intervention. There was also a nationally publicized case in 1984 of a young woman suffering from severe cerebral palsy and arthritis who wished to starve herself to death. The hospital in which she was a patient force-fed her, however, and she petitioned a court to order the hospital to cease the practice. To nobody's surprise, she lost; suicide is still illegal. In 1986, however, she went before another judge, pleading this time only for the right to refuse medical treatment. She won, but the hospital is appealing the decision. The issue is momentous.

An incident during our research on the disabled posed several basic questions bearing on life and death. While my wife, Yolanda, and I were visiting a government office, a young employee who has moderate cerebral palsy wheeled in with tears streaking his face. After he calmed down, he told us that a man from another department down the hall said of him to a companion, "I'd rather be dead." This set in motion a number of queries in my mind: Why did the man say this loudly enough to be overheard? Why was my young acquaintance so stricken by the remark? Why did it upset me so? And, above all, would one really be better off dead? The last is the big question, for to answer it we will also have to ask what constitutes living.

PART I
IN THE BEGINNING

1

SIGNS AND SYMPTOMS

*If the body is not a thing, it is a situation
. . . it is the instrument of our grasp upon
the world, a limiting factor for our
projects.*

— Simone de Beauvoir,
The Second Sex

There is a halcyon period in the life of the middle-class American male that comes at some time between the realization of his ambitions and the start of serious physical decline. It is a time when his earning potential and position in life are at or near their zenith, but before the triple bypass or the discovery of diabetes. Despite this, or perhaps because of it, middle-age ascendancy is usually a time of self-doubt and fear of failing, for it poses the terrible question: Is this all there is? I had little time to ask this question, for my own golden middle age lasted only one week.

On July 1, 1972, I ended a three-year term as chairman of Columbia's Department of Anthropology, a task I detested. The outsider may think of a chairmanship as the pinnacle of an academic career, but this only reveals innocence of the hard underbelly of university life. Let us start with the premise that most

9

professors entered their professions to teach, write, and do research in some subject in which they somehow became interested—all of which is the very antithesis of the administrative role. Research and writing require a certain tranquillity and isolation, as well as substantial blocks of time. Administration offers only continuous interruption, constant involvement in meetings, endless paper shuffling, political bickering, Byzantine gamesmanship, and total immersion in detail work, most of it highly unimportant.

There are few compensations for this. A chairman in most schools receives no tangible rewards beyond a slightly lowered teaching load, and there is generally no extra pay, despite the fact that the hapless incumbent usually keeps on working long after his colleagues have gone about their summer activities. One might expect that a chairman would be treated more generously by the university administration, but in one year I was the only member of my department who did not receive a pay raise—a dean's revenge for my attempt to bring our teaching load into line with that of large departments in other schools.

That particular event is illustrative of the chairman's dilemma. He or she is caught between the contrary expectations of three constituencies: students, fellow faculty members, and the administration. This triple bind reached bizarre proportions in the years 1969–1972, for Columbia had gone through the student strike of spring 1968, only to discover in 1969 that it was going broke. I endured long meetings with graduate students demanding radical reforms (referred to by the unforgivable neologism *restructuring*) and with deans demanding budget cuts. To cap it off, there were student strikes in two of my three years in office. Three years of this kind of pressure left me burned out, fatigued, and eager to get back to my own work. At the end of the day on June 30, 1972, I turned over my duties to the next victim and happily walked away. Yolanda, the children, and I were soon to leave for a month's vacation in Maine, and then there were a couple of books I had been wanting to write. The good times had started, and I thought they would last many years—after all, I was only forty-eight and in perfect health.

On July 7, a week after my emancipation, I threw out a heavy, old air-conditioner, and the next day I noticed that I had a peculiar muscle spasm in my anus. It was a tightness that would not go away and that seemed to have no special relation to bowel function. My first reaction was to shrug it off, for I was brought up in a tradition that treated illness by ignoring it. This attitude was the product of the Depression-era inability to pay doctors, the value of stoicism in the face of pain, and an unwillingness to face up to an unpleasant fact. Most symptoms do go away, most ailments are self-limiting, and I decided to wait it out for a couple of weeks, convinced that I had simply pulled a muscle when foolishly lifting a heavy weight.

The spasm persisted, however, and one afternoon I had difficulty in urinating. I went to see my internist right away, for I already had a premonition that something was very wrong. The visit was reassuring, for it took our family physician only a few minutes to discover that I had an anal fissure, a break in the muscle ring that was repairable by a simple surgical procedure. Since we were about to go on vacation, I decided to put off the surgery until fall. Before going, however, I went to see a surgeon, who tried to break the spasm by injecting Novocain into the sphincter. It was the most painful shot I had ever had, but it worked. The spasm went away—only to return two days later.

The spasm worsened during the summer, and another one developed in my abdominal wall. Upon returning from vacation, I was told by the doctor that the gut twinge probably came from straining during elimination, and he hoped it would clear up after the operation. I liked this kind of cheery advice and straightforward, commonsense diagnosis, for it avoided unpleasant possibilities. Besides, I did have a fissure in the muscle, so it was perfectly logical to assume that it was the cause of the spasm. In October I took a few days off from school and went to the hospital. The *procedure*—the term used by the medical establishment for any sort of intervention in the body of a patient—was simple and I was out in three days. But in the weeks ahead, as the discomfort from the surgery wore off, it became obvious that the spasm was still very much there.

11

The world has little sympathy for middle-aged men with rectal problems, and most people found my complaint amusing. I did, too, until I realized that the operation had done nothing to resolve my condition. I went to a different physician, but as the months went on, slight muscle tensions appeared in other parts of my abdomen, and my new doctor was unable to explain why this was happening to me. The months became two years, and, though the spasms increased, they did so slowly, almost imperceptibly. In any event, they still were not serious problems—or so I convinced myself—and I went about my normal business. About the only change in my routine was that I took up jogging in the naive hope that this would help ease the muscle twinges. Jogging turned out to be so boring and senseless that I soon decided the spasms were preferable; besides, it seemed to make them worse.

Illness and impairment, we are told, are psychological and social conditions as well as somatic problems, and this came home to me vividly. People in good health take their lot, and their bodies, for granted; they can see, hear, eat, make love, and breathe because they have working organs that can do all those things. These organs, and the body itself, are among the foundations upon which we build our sense of who and what we are, and they are the instruments through which we grapple with and create reality. As Simone de Beauvoir wrote, anatomy may not be destiny, but it is indeed an unstated first assumption in all of our enterprises. Each person simply accepts the fact that he has two legs and can walk; he does not think about it or marvel at it any more than he would feel gratitude for the oxygen content of air. These are among the simple existential conditions of life. I will have much more to say about all this, but for now it need only be said that illness negates this lack of awareness of the body in guiding our thoughts and actions. The body no longer can be taken for granted, implicit and axiomatic, for it has become a problem. It no longer is the subject of unconscious assumption, but the object of conscious thought. And so it was that, for the first time in my life, I began to think a great deal about my physical condition. I was becoming quite self-conscious, and in a very unpleasant way.

The rectal operation was my first hospitalization, for I had always enjoyed excellent health. Aside from the usual childhood diseases, the only noteworthy thing that had happened to me was that at the age of four I suddenly lost the use of my left leg. This remains one of my earliest memories. I was playing in the back yard of the house behind ours when my leg became paralyzed. I couldn't move it, and when I tried to get up, I fell. My memory blanks out at this point, and I assume that my mother took me into the house, for my next snippet of recall is of being in bed. The family general practitioner was called, and after a quick examination, he announced it to be "a touch of rheumatism," a bad diagnosis even in 1928. But true to his prediction, I was up and about in two weeks, and I had almost forgotten about the episode until my body began to invade my thoughts in 1972. But even then, I didn't associate it with my current troubles.

The progress of my illness was neither rapid nor dramatic, and its most profound effect was upon my consciousness, my self-awareness, the way I apprehended and constructed the world and my position in it. It was as if a mortgage had been placed on my thinking, as if a great uncertainty, an unspoken contingency, had entered my life. This reaction will sound familiar to anyone who has experienced a serious or chronic illness, but it was totally new to me. My only prior sicknesses had been occasional colds, and I even enjoyed good health during anthropological field trips to the Amazon and Africa. My body gave me no trouble, and I gave it little attention. This was all changed by my altered condition, however, and though I did not brood about my health, my thoughts occasionally would return to it. My problems also introduced a new wedge of anxiety into my general mood. A pall had been cast over my thoughts and into that shadowy area just below thinking, where dwell the things that go bump in the night.

A neurologist once told me that most of his patients came to him with symptoms that were three years old, and I was no exception. As with his other patients, my own delay was the product of misdiagnosis, complicated by my own denial of illness. The operation on the anal sphincter had placed me in a no-man's-land of medical territoriality. My internist deferred to the surgeon, and the surgeon had no idea what ailed me. It was at that point that I went to the new doctor, who proceeded to put me through a battery of tests. There were the obligatory upper and lower GI series, kidney and liver examinations, and a few others that I have forgotten about. All revealed my internal organs to be in perfect health. I expected as much, for I told the physician that the feelings of tightness and discomfort seemed to be on the surface of the abdomen rather than deep inside. The tests confirmed this, and the internist concluded that my problems were probably of psychosomatic origin.

There is no doubt that much physical illness has its origins in the mind, or at least it may be aggravated by psychological processes. The first triumph of psychoanalysis was Freud's cure of a woman with a paralyzed arm, and I had firsthand acquaintance with the often-remarkable success of shamans, or medicine men, in alleviating their patients' symptoms. One of the areas most vulnerable to such ailments is, not surprisingly, the digestive system, and my internist's diagnosis was credible. In any event, his opinion served the purposes of both doctor and patient. Psychosomatic etiology is a handy residual category when standard tests fail to reveal a somatic disorder. How much easier it is to tell the patient, "It's all in your mind," a pronouncement that has the additional virtue of making it, however vaguely, the patient's fault. The annals of medicine are replete with stories of psychosomatic illnesses that proved to be cancer, but there are even more in which the opinion turned out to be accurate. Diagnosis is indeed an art form, but the need for the art is dying today with the tremendous advances being made in scientific examination and testing. Yet what I needed was not a new instrument, but an old-fashioned clinician with plenty of intuition.

The diagnosis was good news for me, because I had begun to suspect that my problems were grave. The idea that I was suffering from a somatized depression meant that a period of some kind of psychotherapy and medication would cure me, a far better prognosis than that for an irreversible disease. The internist prescribed an antidepressant drug and gave me the names of a number of local psychiatrists. A whole arsenal of long-entrenched defenses made me a poor prospect for psychoanalysis, despite the fact that I have always been one of the strongest anthropological advocates of Freudian theory. I used to rebut claims that it was clinically good but theoretically weak by saying, "No, it's questionable treatment but great theory." Nonetheless, I dutifully made an appointment with the first person on the referral list.

The first two sessions were devoted to exploring my physical complaints and my current mood. The usual symptoms of depression seemed to be curiously absent. No, I didn't suffer from insomnia, and no, I had no more difficulty getting up in the morning than I ever did. No, my appetite had not lost its edge, nor had my sex urges. Moreover, I felt no loss of self-esteem, nor was I sad, withdrawn, or despondent; I was not suffering from melancholia. In fact, my most serious psychological problem was a gnawing worry about my health. The analyst pondered my responses and then told me that he had rarely seen a depression so completely rooted in the body, and, more specifically, in its muscle shield. This made it uniquely amenable to orgone therapy, he said, and suggested a course of treatment in an orgone box. "You're a Reichian," I exclaimed, and the doctor hurriedly beat a retreat, saying that we could follow standard psychotherapy if I wished. Wilhelm Reich's theory that the body and psyche receive energy from orgone (a word directly derived from the same root as *orgasm*) radiation, which could be trapped in specially constructed boxes in which the patient sits, had a limited vogue in my younger years, and I was surprised that it still had a following. I had thought it to be nonsense in 1950, and nothing had happened in the interim to change my mind. So ended my adventure in psychiatry.

As my muscle symptoms became more pronounced, Yolanda said that she thought I should see a neurologist. This, of course, is exactly what I had been trying to avoid for months, and when my internist finally gave the same advice, I decided that I could no longer hide behind the happy myth that it was all mental. As a medical specialty, neurology occupied the same position in my mind as oncology, the science of malignancy: It was bad news. I knew little about neurology except that its maladies were often serious and life-threatening, and that many, if not most, were irreversible and incurable. It was, then, with fear and loathing that I made an appointment with a specialist in New Jersey in April 1976.

My initial examination went through the standard neurological workup. The strength of arms and legs was tested, and skin sensitivity was examined. The neurologist jabbed my arms and legs with pins and held a vibrating tuning fork against my feet, asking me to say when I could no longer feel its tremor. I was also given the standard EEG (electroencephalograph) test of brain function. But the most revealing test of all was the simplest. This was the examination for the Babinski reflex, in which a hard object—often a key—is scraped along the sole of the foot. This produces no discernible response in ordinary people, but when it is done to children under two years of age, or to persons with a major impairment of the central nervous system, the big toe curls upward. The doctor ran a key along the sole of my left foot, and my big toe curled upward.

The central cortex, which includes the brain and spinal cord, is a large area, and further tests were needed to localize the trouble. I was referred first to a radiologist for a comprehensive series of X rays of my spine, which revealed nothing abnormal, and then went for a CAT (computer-assisted tomograph) scan of my brain, which was also negative. The next step was a myelogram, a common examination procedure in which a dye is injected into the interior of the spinal column, making the area accessible to X rays. The patient has to remain in the hospital overnight after the test, since violent headaches can result from sitting, standing, or moving about soon afterward. For this reason, and because it is not without risk, the myelogram is not a standard part of physical examinations, and potential threats

16

to the spinal cord can lurk undetected until they become symptomatic.

I checked into the hospital and went immediately to the radiology department, where I was injected with the dye and placed on a tilting table for X-raying. Shots were taken first in a horizontal position, then with my head down, and last with my head elevated. This allowed the dye to flow the full length of the spinal canal, outlining its contents from the cervical to the sacral areas. I went back to my room, but after lunch I had another series of X rays with my head tilted down and my feet up. I was told that there had been a problem with the first set. Later that afternoon, the neurologist came to my room to tell me that they had found an "obstruction" inside the spine, which undoubtedly was the source of my troubles.

The news left me numb, although its full dimensions did not sink in until after I had been fully diagnosed. I immediately told the doctor that I wanted to be transferred to the Neurological Institute at Columbia-Presbyterian Medical Center, where I could receive the care of specialists having extensive experience with spinal cord problems. He agreed, noting that the blockage of the spinal canal was so complete that no dye could flow into the cervical, or neck, area. In order to determine the upper extent of the growth, they would need to do a cisternogram, a procedure in which dye is injected into the top of the spinal canal by a long needle introduced through the side of the neck. The local hospital had no facilities for this, he said, so he arranged for my transfer to the Neurological Institute and referral to a member of its staff.

It was clear to me now that my troubles exceeded my worst fears. My left leg had weakened before going to the hospital, and the irritation of the myelogram dye made me limp for a few days after the test. By this time, too, the muscle spasms had affected much of my torso, front and back, and new symptoms were appearing at an increasing rate. Now that I knew the cause, I began to have a retrospective understanding of some of the symptoms. I had lost my balance and fallen from a stepladder a few weeks before the obstruction was discovered, and an occasional slight stagger made my friends wonder if I had taken to drink. My sense of coordination and balance had been deteriorating in a process that was so slow and subtle I barely had noticed it.

I also had tingling sensations in my left foot, which I attributed to my two-and-a-half-pack-a-day smoking habit. The tingling occurred every morning at breakfast while drinking coffee, shortly after lighting up my first cigarette of the day. After a few deep drags, I would feel a sensation in my left foot much as if it had "gone to sleep." To confirm the hypothesis that nicotine was causing my capillaries to contract, I waited two agonizing hours one morning before smoking, and the tingling happened at the end of my second inhalation. It was a splendid scientific experiment, but I neglected to think about why it happened to the left foot and not the right. In fact, my ailment had so slowed the blood flow to the left foot that the additional constriction caused by a bit of nicotine was enough to bring it to the threshold of awareness. But who would ever have suspected that a little tingle—or a stiff gut muscle—could be a symptom of a major neurological affliction? The tingling foot succeeded, however, in doing what the Surgeon General's Report had not done: It convinced me that I should give up smoking, which I did, cold turkey. It made for a bad couple of months, leaving me convinced that giving up tobacco is harder than going off alcohol.

I went to the Neurological Institute with very mixed feelings. On one hand, I was optimistic that I would find the best care available and that my condition would be treated successfully. On the other hand, I was not looking forward to the means they would use to reach that happy outcome—starting with the cisternogram. But what depressed me above all else was the realization that I had lost my freedom, that I was to be an occasional prisoner of hospitals for some time to come, that my future was under the control of the medical establishment. I had fallen into a vast web, a trap from which I might never become extricated, and I had no choice but to relinquish even the illusion of free will and submit to a new and alien order. My sense of entrapment went far beyond normal forebodings about hospitalization, for beneath my fond hope for a cure I suspected that I could be headed into chronic illness. I was not simply confronting an unpleasant two or three weeks in the hospital, but a new way of life, a career of being "sick."

Talcott Parsons, the dominant figure and chief guru of mid-twentieth-century sociology, wrote several papers on sickness as a social role,[1] an area that had been pioneered by the anthropologist David M. Schneider in a 1947 paper on the social uses, and abuses, of illness in the U.S. Army.[2] The Parsons paper was written in the cumbersome language that has blighted sociology, but when properly translated, it said only what is common knowledge to everybody who has ever been sick: A person's ordinary social roles—mother, father, lawyer, baker, student, and so forth—all become temporarily suspended when he or she falls ill. The individual becomes a "sick person," which relieves him or her of some or all of the ordinary obligations, depending on the severity of the illness.

The suspension of his other duties does not mean that the person playing the sick role has none at all. Quite the contrary; he is saddled with one big obligation: He must make every effort to get well again. In our own doctor-ridden culture, this means that he must seek medical advice; he must take his medicine and follow the doctor's orders. This expectation mandates the proper role of the sick as one of passivity. The sick person is excused from work or school, household duties are suspended or at least limited, and connubial relations may be put on ice. But in return, he must devote full time to getting better.

There are rules for being sick. If one is only slightly ill, he or she may be criticized for taking too many liberties from ordinary duties. Thus, soldiers who showed up for sick call with only minor complaints were considered "goldbricks" in Schneider's paper. Anyone who does this too often acquires a reputation as either a shirker or a hypochondriac. We treat the former with contempt or punishment and the latter with derision. At the other extreme is the heroic stance, adopted by the person who ignores pain and suffering and goes about his or her ordinary business. Such types, however, receive only limited honors. The mother who cares for children and household despite serious illness is admired as long as she makes a good recovery. If her condition worsens, she "has brought it on herself," and those who formerly praised her courage will say that "a woman with three young children has an obligation to take care of herself." The mother thus moves from heroine to child neglecter and is regarded as the cause of her own condition. She has violated the first commandment of sickness: Get well.

As with all other social roles, a person can succeed or fail at sickness. A key rule for being a successful sick person is: Don't complain! The person who smiles and jokes while in obvious physical misery is honored by all. Doctors and nurses are especially appreciative of this kind of patient, for he usually follows orders and seldom files malpractice suits. Hospital visitors also value cheeriness, and the sick person soon finds that he is expected to amuse them, and thus relieve their guilt at being well. These are front-area, or on-stage, performances—to use sociologist Erving Goffman's celebrated theatrical metaphor for social interaction.[3] The backstage behavior may be totally different, however, and the public hero may become a whiner at home. The bad patient is either tyrannical or a crybaby, or both at once. But above all, the bad patient is one who does not follow orders. There are, then, social skills in sickness.

Contrary to popular fears, the overwhelming majority of people who enter hospitals come out not only alive, but also in better health than when they entered. What one loses in the hospital is not life but freedom of choice, for the patient must submit to the requirements and routines of the institution. When an individual is additionally removed from his normal habitat and is placed under custodial care, his passivity is complete. He is shorn of all other social roles and is regarded as one sick body among many. Some attention is given to age, sex, and social standing, but this has less effect on care than might be imagined. In general, hospitals are more democratic toward their patients than is society at large; they withhold honor with an even hand. This is particularly irritating to medical people when they become patients and find themselves treated as minors in the same establishments in which they normally reign supreme. The hospital inmate soon learns that he must conform to the routine imposed by the establishment. If dinner is scheduled for 4:30 P.M., as it was on a floor in which I once spent two months, then that's when you eat. And if your bowels don't move often enough to suit the nursing staff, laxatives are the answer. The infamous routine that demands that all temperatures be taken at 6:00 A.M. is well known to all who have been patients. I even spent five weeks on one floor where I was bathed at 5:30 every morning because the daytime nurses were too busy to do it.

The patient conforms in still other ways. There is a chain of authority that goes down from attending physician to nurse to patient, with obedience expected at each level. All of this is understandable, however onerous. Given its problems of size and complexity, the average general hospital must work according to a thoroughly "rationalized" system. There is an elaborate division of labor, a meticulous allocation of responsibilities, and a careful scheduling of activities. The hospital has all the features of a bureaucracy, and, like bureaucracies everywhere, it both breeds and feeds on impersonality. It should be cautioned, however, that the "rationality" of a bureaucracy refers only to the attempt to impose order, not to its actual operation. To the contrary, as the sociologist Robert Jackall has demonstrated brilliantly, the practical results of these careful designs are often exercises in madness, and hospitals are case studies of the awesome disparities between theory and practice. They are places in which both Florence Nightingale and Nurse Ratched of *One Flew Over the Cuckoo's Nest* could feel at home.

To a certain degree, the hospital is a "total institution," a custodial place within which the inmates live out all aspects of their lives; it is an island of social relations.[4] The totality is not as great as that of the military or the prison, but in long-term care facilities, such as mental hospitals and physical rehabilitation centers, it comes close. The truly closed-off total institution generally tries to expunge prior identities and to make the individual assume one that is imposed on him by authority. Prisons and the military give the newly inducted close haircuts and a

21

number. The hair grows back, but the numerical identity is more enduring; I once absentmindedly listed my Navy serial number instead of my Social Security number—twenty-five years after my discharge. The purpose of all that is to make the individual forget that he is somebody's son or husband and think of himself instead as a soldier or convict. The Navy had brilliant success in my own case, and my early professorial career was colored by an uneasy feeling that I was a seaman who had gotten on the wrong ship. I had been so thoroughly imbued with the sailor role that I had a hard time adjusting to the academic style.

The hospital requires that the inmate think of himself primarily as a patient, for this is a condition of conformity and subservience. This allows the medical staff to treat the patient with a degree of distance and dispassion, to view him or her as a case rather than as a person. The patient may not like this treatment, this evenhanded pleasantness, but he is in fact a co-conspirator in the process; after all, his other roles and duties have already been suspended. This is all common-sense knowledge to most people, and it has been the subject of countless scholarly papers, but it bears repeating because I knew exactly what kind of social morass I was getting into.

The Neurological Institute is a part of the Columbia-Presbyterian Medical Center and a teaching hospital of the Columbia College of Physicians and Surgeons. It is somewhat different from most hospitals I have visited. With only 100 beds, it is smaller; with long-term patients, there is less turnover. I suspect, too, that the longer stays mean there are fewer visitors. In any event, the Institute is quieter than other parts of the hospital complex, and its small size makes for more relaxed rules. The nurses and aides also become better acquainted with the patients, so relations are less impersonal. I was never asked, "How are *we* today?"

These positive aspects compensated for the archaic condition of the building and its furnishings. When I entered in May 1976, my room did not even have the usual lighting behind the bed. Instead, there was a rickety floor lamp. The bed was the old, manually operated type, and the furniture was on wheels, offering inadequate support for unsteady neurological patients. Except for the more expensive private rooms, beds were shoehorned into small spaces, leaving practically no room for guests, and "air-conditioning" came from antique fans. Indeed, the condition of its building was so bad that the Institute was threatened with loss of accreditation a few years ago. Only the hospital's decision to erect a new building that would also house the Institute saved it from being shut down.

The heart of the Neurological Institute is not its sixty-year-old building, however, but a superb medical staff. The first thing I learned from my new clinical neurologist was that a cisternogram would not be necessary. The X-ray plates sent over from the specialist in New Jersey showed clearly that the myelogram dye *had* filtered above the obstruction, the outlines of which were visible from top to bottom. That was the good news. The bad news was that the obstruction was a tumor extending from the second cervical vertebra to the eighth thoracic vertebra, or from the top of the neck to midchest—half the length of the spinal column. Fortunately, it was thin in its upper reaches and was not pressing against the cord—yet. It was, however, growing steadily, though slowly, and if left unchecked, it threatened to destroy my entire central nervous system, except for the brain. On the bright side, the doctor observed that it was an ependymoma, a benign tumor that grows on the ependyma, or lining, of the spinal cord. I listened to him in stunned silence, for my nicely ordered and rather comfortable existence was falling apart before my eyes.

The probability that the tumor was nonmalignant may sound hopeful, but neural tumors present a problem different from those in the trunk of the body. A benign intestinal tumor can grow to a large size without causing great or lasting damage, and it usually can be removed successfully, along with neighboring tissue. Neural tumors, however, grow within the restricted confines of the skull or the narrow constraints of the spinal canal. When they reach a certain size, they exert pressure on brain or cord, and nerve cells die when they are pressed. Moreover, one of the more distinctive characteristics of nerve tissue is that, once dead, it never regenerates. Restoration of function can only be effected by the body's use of alternative neural pathways, and the possibilities for this are limited. Most neural damage is, therefore, permanent and irreversible. Benign brain tumors can blind people, cripple them, drive them to madness, or kill them. And a benign spinal cord tumor can totally and irrevocably paralyze a person from neck to feet, eventually sentencing him to a respirator. On the other hand, the doctor added, ependymomas seldom kill people; you can live for years with one. I thought a moment, then asked, "Is that good news or bad news?"

It is difficult now to recollect my state of mind at the time, but I took the diagnosis very calmly. Most people neither rage nor weep on hearing such dismal tidings, nor do they become despondent or suicidal. Many are unable to assimilate the full meaning all at once and may sit dazed for hours until the heavy weight of truth finally sinks in. It took time for me to realize the significance of what I had heard, enough time for my psychic defense system to become mobilized and throw up a wall between me and an unpalatable reality. I do remember that I clutched at straws. Perhaps it was easily removable by surgery? The neurologist was very guarded and cautious in his reply to my question, and he made it clear that tumors of this type are difficult to excise. He obviously did not want to raise my hopes too high, but beneath his reserve I detected a certain pessimism. One must remember, the doctor said, that the tumor has grown half the length of the spinal cord and that it has been growing for a long time. How long? For many years; it might even be congenital. It was then that my childhood episode of paralysis fell into place.

The neurologist warned me not to jump to any premature conclusions, for they still knew very little about the tumor. No prognosis was possible until after an entire battery of tests. I accepted this wait-and-see attitude very easily. It was the only alternative to contemplating the worst-possible-case outcome of the disease. I once asked the neurologist how bad it could get, and, with a pained expression, he answered, "Do you really want to know?" I didn't.

I gradually learned to live day by day, to block from my consciousness any thoughts about the final outcome of the illness, to repress from awareness any vision of the unthinkable. I have maintained this perspective for the past ten years. In the many stages of my progressive debility, I have guarded against meditation about what would come next. And as I have confronted every affront to my body, whether inflicted by nature or by medicine, I have lived for the moment. Lest this be misunderstood as an attempt to escape the truth, I did (and still do) know what could happen to me, but I knew also that there wasn't a damned thing I could do about it. It is the worst kind of human foolishness to worry over things that are inevitable. It's like fretting over death; you can spoil your life that way.

People mistakenly see either great courage or unabashed bluff in the fact that I have gone calmly and cheerfully about my business on the eves of surgery and that I face the future with outward equanimity. It is, however, neither courage nor bluff, only self-anesthesis of emotion and stream of consciousness, a long and carefully inculcated gift for foreclosing tomorrow and repressing today. I enjoy a healthy share of physical cowardice—the only sensible stance for a person who grew up as one of the smallest boys in his class—but I can say truthfully that I felt no fear upon hearing the diagnosis, other than anticipation of a dismal future. It was as though the real me were off to one side, watching all this happen to someone else. My strongest response was one of amazement and awe. "I'll be a son of a bitch," I said, which was almost exactly the way I had reacted to naval warfare more than three decades earlier—I was less a combatant than a spectator. This wasn't bravery, just an early manifestation of the Disembodied Self, which would reassert itself strongly in the coming years of disability. This technique was not just my own personal dodge, but a common reaction of people under severe stress. The Vienna State Opera, for example, performed according to schedule on the night before the Soviet Army entered the city in 1945, and studies show that survivors of natural disasters often try to fall back quickly into prosaic routines. People live in the present when the future is fraught with danger and complete uncertainty, as mine most certainly was—and is.

The theme of "one day at a time" has become such a cliché in the United States that the phrase has even been used as the title of a television situation comedy, perhaps the ultimate banality. In my case, however, I am talking about a frame of reference that goes far beyond the motto of "sufficient unto the day the evil thereof" or simple pleasure seeking and invades our own perception of our personal histories. My past is divided radically into two parts: pre-wheelchair and post-wheelchair. I think of the pre-illness years as a golden age and the recent period a time of bad auspices, gloomy auguries, and shattered expectations. My history is no longer smooth and linear, but bisected and polarized. And my long-range future does not really exist.

Now in 1986, at the age of sixty-two, I give no thought to retirement, although I will be eligible for full pension rights in three years. The reason for this is quite simple—I do not expect to live to be sixty-five. This may well turn out to be a false expectation, just as was my belief in 1976, after learning of the dimensions of my problems, that I would not make it to 1980. But no matter how wrong my predictions have been, my attitude remains the same. I hesitated a long while before applying in 1980 for a two-year research grant because I was concerned that my health could interfere with completion of the program. Most of my plans cover no more than a year. Beyond that, the future appears as a blank, impenetrable wall. One cheery soul said to me, "You should think of this as the first day of the rest of your life." "Okay," I replied, "ten . . . nine . . . eight . . . seven. . . ." The odd part of all this is that I do not find it very depressing. That's just the way things are.

The key to a successful strategy of deferring thoughts of tomorrow is a well-tuned repression mechanism, the ability to become detached from one's emotions, to benumb the inroads of fear. This is not an easily acquired gift—or curse, depending on one's viewpoint. My own defense apparatus was the product of growing up Irish Catholic in the 1930s and 1940s. Our family had graduated from lace-curtain, but working-class, Irish to solid middle class during the 1920s, only to crash back into poverty during the Depression. My father had been an advertising salesman for a yachting magazine and was doing well enough to buy a house in Rockaway Beach, a seaside section of Queens, New York. The life of a traveling salesman during the hard-drinking Prohibition Age, however, pushed him over the edge of conviviality into alcoholism, a state for which he had been long prepared by the doting and permissive love of his mother. Between his drinking and the fact that yachting had become moribund by 1931, he found himself among the one-third of a nation without a job, and, except for sporadic WPA work, he remained unemployed for the next ten years.

Lost in alcoholism, depression, and self-pity, my father became increasingly removed from society and family, and as my two brothers, my sister, and I began to realize what had happened to us, we became bitterly resentful of him. We saw him as the cause of our poverty and a source of shame and embarrassment. After about the mid-1930s, our friends rarely entered our house. I still remember him as an isolated, broken man who showed little love for his children, but who never punished us either. I don't think he cared enough or was strong enough to impose discipline. I never once had a prolonged or serious conversation with him, but I did not think this unusual, for at that time there was far less communication between parents and children than there is today, despite recent nonsense about a "generation gap."

The only source of warmth in this cold and straitened setting was our mother, a good and loving woman who fought mightily to protect her four children and hold the family together. My father had gone from one strong woman to another when he married, shifting dependencies and perpetuating weakness, continuing the vicious circle so often found in the Irish family. Then, in 1935, my mother contracted breast cancer, grew steadily sicker and weaker, and died in 1937, her end quickened by childbirth. The center of the family, the one person who had given us both direction and love, had gone, and we all faced a bleak future. I wept bitterly at first, but by the time the funeral was over, my tears had stopped. I have not cried since, although I have grieved deeply many times. My father's mother took over the housekeeping and a bit of the maternal role, and we five siblings still had each other. But we had learned a terrible lesson: Love invites loss.

The years of impoverishment, the emotional absence of my father, and the death of my mother had taught me that in order to survive, I had to suppress fear and sorrow. This fitted in well with the boyhood ethic of the time, a cardinal rule of which was: Never cry. Boys were taught to tough it out, to guard their emotions, to avoid displays of tenderness. It left me unable to vent grief at the times when my four siblings died, but it has steeled me against fear and self-pity in times of trouble.

It might be the most valuable lesson I ever learned, for it got me through a war. On the small warship on which I served, laughter was almost the only emotion that could be expressed. Aggression among crew members had to be squelched because we lived in such close quarters, and any show of fear was absolutely forbidden because it is contagious. Our laughter sometimes grew macabre. On one occasion, an American fighter pilot had the poor judgment to chase a Japanese suicide plane over our invasion fleet, and both were shot down by naval gunners. One of our deckhands the next day painted an American flag on the side of the wheelhouse, along with the Japanese ones. The captain was not amused, but the crew thought it was hilarious. Compassion and empathy had little survival value at the time, and it was only through my marriage to Yolanda and raising our two children that I was able to resurrect my ability to love.

This kind of stoicism, so common in the Depression–World War II generation, is out of step with the contemporary pressure to "be in touch with one's own emotions," an expression that I consider part psychobabble and part narcissism. We now live in an age in which men are urged, usually by women, to show their emotions and, above all, develop the ability to weep. Paradoxically, many professional women are terrified by the possibility that they might break into tears in public when under stress, an implicit recognition that the male attitude still prevails in the marketplace. This kind of repression is bought at considerable emotional cost, but it has its positive uses. Some fears and sentiments are better left unstated, and those that I harbored as I entered the hospital in 1976 were among them. What I refused to contemplate was the progressive and total destruction of my body, the reduction of all volition to quietude, the entombment of my mind in inert protoplasm.

The maximum growth of the tumor, my neurologist informed me, was at the top of the thoracic section of the spine, more or less between the shoulder blades. "Remarkable," I exclaimed, "that's exactly where the Mundurucu believe the soul is located!" I explained to the puzzled doctor that the Mundurucu are an Indian tribe in the Amazon region of Brazil that Yolanda and I had studied from 1952 to 1953. In their view of things, soul loss was a major source of illness, but the doctor was not impressed by my notion that perhaps a good shaman could cure my ailment and even reveal who had inflicted it upon me. The coincidence, however, triggered a train of my own thoughts about the relation between primitive and modern conceptions of disease—and, ultimately, about the ways we impose mental forms on the world around us.

Many primitive peoples trace the source of disease to violations of order. Navaho curing practices seek to restore the sick person and the community into balance with the cosmos, and Azande diviners in the Central African Republic and Sudan usually locate the origin of their patients' illnesses in failures to observe ritual protocol, such as the breaking of a food taboo. In much the same fashion, a Mundurucu may bring the wrath of game-animal spirits down upon himself by failure to maintain reserve and decorum while eating meat, or a mother may cause her child to fall ill by eating a scaleless fish while still nursing. Some of these regulations serve ecological or health purposes, but most symbolically counterpose order and health against disorder and illness. Pragmatically, the beliefs explain illness to the

29

patient and his family, and thus reduce anxiety while making the ailments amenable to healing. On another level, however, they tell us that the social and natural orders are delicate and prescribe a series of compulsive acts for the maintenance of their ever-precarious balance. What is it, then, that makes social equilibrium so threatened that it must be achieved by resorting to the supernatural?

The worlds created by the human imagination are far more coherent and structured than the real social systems in which we live, and the mental constructs by which we make sense of society are only loosely related (sometimes inversely) to what is really going on. We take these conventional views of our social system as matter-of-fact, true representations of social reality, but they are socially constructed realities, human artifices whose purpose is to perpetuate society, not clarify it. And they do so, not by casting a clear light on social life, but by rendering it opaque, even mystical. Because of this, the collective illusions by which we live are vulnerable, fragile, transient. The social order is in good part a mental order, and all disturbances of society involve perturbation of the mind, and vice versa. This must be understood in order to appreciate the full significance of the relationship among tumors, paralysis, and culture.

Among the Mundurucu, and many other South American Indian societies, the most serious threats to health are believed to arise from violations of the social order, specifically from the activities of malevolent and alienated individuals, people who, say the Mundurucu, "are angry at everybody." Most diseases, the Mundurucu believe, are caused by evil shamans who manufacture supernaturally charged objects, called *caushi*, which enter a person's body and cause illness. A good shaman then must be called to remove the *caushi* by massage and suction, and then to heal the damage through a prescription of herbal remedies. When many people in a village become seriously sick, a powerful shaman must identify the sorcerer, who subsequently is executed. The condemnation and execution are done by collective action of the entire community, which is thereby restored to social order and physical well-being.

The sorcerer and his magic are a parable of the hold of disease upon the human imagination and, conversely, of the imagination's use of disease in symbolizing the problem of order. The cause of the illness is an alien object *in* the body but not *of* the body, an object that cannot be seen by ordinary people, and that slowly destroys its victim from within. This is the metaphor of the malignant tumor in our own society, but it can be extended to the almost equally destructive benign tumors of the central cortex. Susan Sontag, in her essay *Illness as Metaphor*,[5] sees cancer as a sharp deviation from the natural order, in that malignant cells multiply and grow without limit. Moreover, their very structure is radically different from normal animal tissue. Sontag calls cancer "the disease of the Other," and in a fine flash of insight, she compares it to the classic science-fiction movie *Invasion of the Body Snatchers*. Film fans will remember that two motion pictures have been released under this title (the first was the best), and the theme has been copied many times. The story seems to have a primordial appeal. In the films, aliens from outer space took over the bodies and identities of earthlings, and so effective was their silent infiltration that even the husbands and wives of those usurped were duped. They only knew that their mates had not seemed "quite themselves" lately. The body-snatcher theme is kept alive today by those wonderful folks who believe in flying saucers and a government plot to deny their existence. This is a classic fantasy of paranoia: Beneath surface appearances of normality, a destructive force is quietly gnawing away at the foundations of society or the self— and an anti-order is being established in its place. This is equally true of cancer, as Sontag has observed: "Cancer is now in the service of a simplistic view of the world that can turn paranoid."[6]

The disturbance of natural order by the intrusion of an alien object in the body is one of humanity's oldest and most pervasive explanations of disease, and I suspect that the intensity of our horror for tumorous growths is an expression of these ancient fears. Sontag has observed that the paranoia carries over into public life. The great hunt for hidden Communists in the

1950s has been compared by any number of writers to the Salem witch trials, and anthropologists have made similar observations about the beliefs and practices of sorcery in simple societies. Just like the invisible malignant objects he manufactures, the Mundurucu sorcerer is thought to operate covertly, maintaining his everyday demeanor and mask of sociability while spreading disease and suffering. Killing the accused restores the body politic, and at the same time purges the populace's fears, envies, and secret hostilities, which have been projected onto his person. The good shaman brings back life and normal function to the body and the community; the evil shaman creates death and disorder—which are really the same thing.

Tumors, as Sontag wrote, excite paranoia, that most fascinating of the psychoses, for there is often only a thin line separating the paranoid from the genius. It is also a very human disturbance, for all of us exhibit attitudes that are consistent with paranoia, albeit in a minor key. One of its most characteristic delusions is that we live in a rigidly deterministic universe. This order is often found beneath surface appearances—there is, so to speak, a hidden agenda, a plot against him that the paranoid seeks to unravel. Such an underground is also, however, a common assumption of the social sciences, which seek to find "latent functions" of social practices—that is, the hidden purposes they serve. Where real paranoids differ is that order is magnified and radically separated from reality. It is the idea of order and causality driven beyond its limits—and thus transformed into madness.

Students of human nature have attributed a large number of urges, impulses, drives, or whatever to the natural endowment of our species, giving an uneasy feeling that the essential human animal is no more than a collage of instincts; most of these bear a suspicious resemblance to current political ideologies, all of them reactionary. Given the extensive inventory of human "innate traits" postulated by this faulty genetics, it is surprising to learn that the old Aristotelian notion that man is a rational creature has only returned to vogue in the last thirty years or so, in the "structuralist" theory of Claude Lévi-Strauss. The core of structuralism is the premise that the human brain has an innate structure, or organization, and that the panhuman propensity for thinking in binary, opposed categories derives from that structure, not from the imposition of culture on a totally malleable mind. The thinking of all humans is much the same, according to Lévi-Strauss: We sort out our perceptions into opposing categories, which come to constitute a framework for our apprehension and understanding of the world. The theory goes on to hypothesize that the mind seeks to project this natural dualism upon culture, and these mental structures are coded into mythologies and marriage systems alike. There is, then, a need for order in all humans that impels us to search for systemic coherence in both nature and society and, when we can find none, to invent it.

Whether the human brain works according to Lévi-Strauss's theory can remain an open question, but it is an empirical fact that the mind seeks to impose systems of some kind of order upon all it surveys. It is a property of all peoples and all cultures, but I doubt that it is a biologically inherited tendency. I would suggest instead that it derives secondarily from our deepest biological urge, the instinct for self-preservation. We look for order because it makes predictability possible, and we seek predictability to avoid danger in an essentially perilous world. Our fondest illusion is that we can reduce risk by making the unanticipated predictable and by exerting human control over the contingencies of life. This can be done by ritual or by engineering; both impose a human order upon formlessness.

Whether or not our structured images of the world around us correspond to an external reality, the predication of an order is necessary for intelligent creatures. It allows us to operate in a cockeyed world and to find meanings for our actions and lives in a milieu utterly devoid of absolute meanings, either natural or God-given. Tumors are a metaphoric breach of order, an assault upon both flesh and thought, and I had a big one right in the spot where my soul would be, if I were a Mundurucu. And like all neural tumors, it infringed upon both symbolic coherence and real systems of motor control. This is manifest in the spastic movement of people with cerebral palsy or, in my own case, in the slow but steady atrophy of both my voluntary motor capability and the regulation of my autonomic functions. Signals from my lower body could no longer reach my brain for processing and response. My body was no longer governed by a higher system of control; I had a dis-order in the most literal sense. I was drifting into motionless inertia. I was on a journey into entropy. *

---

* Entropy (*Webster's Unabridged Dictionary*): A measure of the degree of disorder in a substance or a system; entropy always increases and available energy diminishes in a closed system, such as the universe.

2

THE ROAD TO ENTROPY

Civilization can be described as a prodigiously complicated mechanism: tempting as it would be to regard it as our universe's best hope of survival, its true function is to produce what physicists call entropy: inertia, that is to say. . . . "Entropology," not anthropology, should be the word for the discipline that devotes itself to the study of this process of disintegration in its most highly evolved forms.
—Claude Lévi-Strauss,
Tristes Tropiques

My first two weeks in the hospital were spent in taking tests. The laboratory had a Draculan appetite for my blood, which was drawn every other day, or so it seemed, and I was again X-rayed from every conceivable angle. In addition to these routine examinations, I was given two EMGs (electromyographs), which measure the extent of neural damage by shooting electric currents through certain nerves. This is done by inserting small, Teflon-coated electrode needles into the body and reading conductivity responses. The electricity used is of such low voltage that it is not perceptible, and it is only the needles that bother the patient. Although they are very thin and only penetrate one-fourth inch or so, they make up in number for their lack of size. Each of my hands required a dozen readings, and my arms and shoulders took a dozen more. And to make the process even more uncomfortable, the examining neurologist occasionally

35

wiggled the electrodes. I remember vividly her saying to me, as she joggled an electrode buried in the ball of my right thumb, "You must relax, Mr. Murphy."

Much more memorable than the EMGs was the arteriogram, in which a dye is injected into an artery and an X ray is taken immediately of a portion of the vascular system—in my case, that of the spinal cord. My first hint that this was to be an unusual procedure came when I was given a preliminary shot of Demerol, a powerful sedative, and the second came when a glucose drip was inserted into a vein in my foot. After that, my body and feet were strapped down, my head was taped to the table, and my arms were extended sideways and bound to boards. A long needle was then inserted into the main artery of my arm with the help of a fluoroscope. To make sure that the needle was in the correct position, the radiologist dropped a trace of dye into the artery, warning me first that it would give me a sensation of warmth, which it did. The trial run did not prepare me, however, for the full injection of dye, which made me feel as if my body temperature had been raised hundreds of degrees. It was much like being roasted alive in a microwave oven. The radiologist had told me that the sensation would not last long, and I counted off twelve of the longest seconds in my life. After a second injection, it was all over.

As I was wheeled back to my room, I began to realize that the entire episode was very reminiscent of something. It was a sort of crucifixion—my arms stretched out and pierced, my feet tied down and skewered by the IV. I remembered Christ's lament from the Mass: "They have pierced my hands and my feet, and they have numbered all my bones." The thought that I had somehow embarked on an imitation of Christ both annoyed and intrigued me. Despite years of presumed liberation from the Catholic Church, an extremely threatening experience had driven me back into its forms, and I had found momentary refuge from the troubles of my history through shelter in timeless symbols. For a few minutes, I felt kinship with medieval flagellants and penitents making their way on hands and knees to the Irish pilgrimage church of Knock. My chagrin at this lapse was soon replaced, however, by an anthropological interest in it, for we can no more escape our pasts than can the most primitive of our subjects. Alfred Kroeber, the dean of American anthropologists until his death in 1960, once asked me, "I suppose, Murphy, that you have left the Church?" "Yes," I replied, "but the Church hasn't left me." And what stayed with me was not the external beliefs and practices of Catholicism—for not even at the worst points in my illness have I had a moment's conscious recourse to religion—but its dramaturgy. God was dead in my mind, but guilt and atonement were alive and well. We will come this way again.

Beyond the testing, much of my time in the hospital was spent in answering questions. Patients with interesting diseases in teaching hospitals are much like eyewitnesses to crimes: Every new person who comes on the case has to hear the story. I thought seriously of typing it all, from first symptom to most recent, and giving a copy to each new medical student or resident, but I knew that asking good questions was a goal of their training, so instead I helped them out. Each arrival also did the standard neurological workup, complete with pin-sticking, mallets, and tuning forks. After three weeks of this, and of being a regular stop on the weekly grand rounds, I began to think the university owed me a salary for summer teaching. But the process was an educational experience for me too, such as on the day that a recently retired clinical neurologist asked if he could use my body to demonstrate a standard workup to a young resident. He went through the entire process for an hour, explaining to his student the purpose of each test and the meaning of each of my responses, both verbal and reflexive. By the end, I knew a lot more about my malady, and I also knew that he could have diagnosed my illness without a CAT scan or even a myelogram.

After all the tests had been finished, I was visited by a neurosurgeon who, I thought, would discuss surgery with me. Instead, he explained that my examination results had been reviewed, and the medical staff had concluded that I was a poor prospect for surgery. The tumor had been established for so long that it had wrapped itself around the spinal cord, making it difficult to excise without causing serious damage to the cord itself. Moreover, some of the vessels that supplied the spinal cord with blood ran through the tumor. They could not be tied off during surgery because the deprivation of its blood supply would kill the cord. It was a no-win situation; every surgical step would pose a risk to the cord greater than that of the tumor. The surgery could produce immediate and catastrophic damage, while the tumor at least worked its harm slowly.

My basic problem arose from the fact that I was fifty-two years old when the tumor was discovered. Many ependymomas become symptomatic in childhood, when a surgeon can perform a simple laminectomy, or opening of the spinal canal, and peel off the tumor with one motion. Using surgical techniques perfected in the last five years or so, surgeons now remove children's spinal cord tumors with the aid of lasers and ultrasound. The value of early diagnosis was brought home to me clearly when a friend visited me in the hospital. Upon hearing what was wrong with me, he took off his shirt and showed me a scar on the upper end of his spine, the residue of an operation thirty years earlier, when he was twenty. He had had the rare good fortune to bring his complaints to the attention of a general practitioner who knew something about neurology. If my "rheumatism" had been diagnosed correctly when I was a child, I might still be walking. On the other hand, given the state of the art of surgery in 1928, I might also be long dead. This only proves that it simply doesn't pay to play "might have been" in matters like this. On many other occasions I have wondered whether I would have been better off or worse if one procedure or another had or had not been done, but there is no way of knowing the answers. In any case, I knew that I owed it to my family, if not to myself, to give everything a try.

I had had high hopes for a successful operation, and now they were thoroughly dampened. But I was told that the situation was far from hopeless. Instead of surgery, the doctors recommended a course of cobalt therapy because the cell structure of neural tumors is sufficiently different from normal cells to make them vulnerable to radioactive bombardment. In response to my questions about radiotherapy, my neurologist warned that normal cells also suffer some damage in the process, but not as much as tumorous growths receive—it would be a trade-off. He was careful to make no promises or prognoses for the treatment, but nonetheless, I now attached my hopes to radiation.

The radiotherapy department is located in a sub-basement of the hospital, well out of sight of the general public. Most of the clientele, both inpatients and outpatients, were being treated for cancer, which lent them a lugubrious tone—in marked contrast to the professional cheeriness of the staff. It was a sad place. The mother of one little boy, whose hairlessness revealed the inked lines that guided the radiation, said, "He never laughs anymore." Neither did most of the habitués. This made my daily visits depressing experiences, and each departure was an escape.

The medical personnel of the department included an unusually large number of Orientals. This wasn't too surprising, for many minority practitioners undertake specialties like this that are a step removed from the public, and thus from racial prejudice. My only uneasiness about this arose when I couldn't understand the English of the Japanese doctor who was marking my back for radiation. I assume that he got it right, since, for better or for worse, my spine was indeed radiated from one end to the other.

The cobalt therapy began when I was still in the hospital and went on for seven weeks, four of them when I was an outpatient. The routine was monotonous. At about the same time every weekday, a "transporter" would come to my hospital room with a wheelchair and push me through the tunnel system to radiology. When I was finished, another man would push me back. When my name was called, I would walk into one of the rooms and lie face down on a stretcher, which was then positioned under the machine. The technician would go back to her control panel, from which she could watch me on closed-circuit TV, call out that I should keep my body still, and then activate the device. It would buzz for about ninety seconds, and then I would be ready to go back to my room. Other than these excursions, I had nothing at all to do in the hospital, and I whiled away my time reading mysteries, talking to visitors, and watching TV—I still don't know why I was kept in the hospital for the first three weeks of radiation. I tried to do some serious reading, or at least to read doctoral dissertations, but there were too many interruptions by visitors, nurses, meal trays, and so on. Besides, I had fallen completely into the pattern of role suspension, the lot of the sick.

The weekdays in the hospital were easygoing and the weekends were blanks. Most therapists and technicians work from Monday to Friday, and all the attending physicians take the weekend off, leaving the hospital in the medical custody of the residents, who are also its nighttime caretakers. There being no reason to stay in the hospital on Saturday and Sunday, ambulatory patients are given weekend "passes," much like military people. In fact, my returns on Sunday evenings were marked by the identical sense of reimprisonment and depression I once felt as a nineteen-year-old returning on Sunday nights to naval bases. Again the hospital showed its remarkable ability to reduce its inmates to juvenile status and to reawaken past states of mind, forgotten but unforsaken.

The cobalt rays were aimed at my spine but of course they kept on going, hitting my digestive tract from throat to intestine. As the therapy progressed, all my food began to have the same terrible taste, and my appetite ebbed. I forced down food past waves of nausea, knowing that if I stopped eating, I would become even sicker. The radiation also left me feeling tired and listless, and the hospital and its food began to pall on me badly. My release to outpatient status after six weeks of hospitalization—from early May to late June of 1976—overjoyed me, and not even the daily trips for radiotherapy tempered my sense of release. The radiation sickness continued, however, until a few weeks after the end of the therapy, and my life did not return to near normal till mid-August.

During the first few weeks of the radiation, I had days when I felt better, which convinced me that the therapy was helping. But these hopeful signs faded as the weeks passed, and by the end of the radiation, I was, if anything, a bit worse off than

40

before. By the time I returned to teach classes in September, my left leg had weakened so much that I limped perceptibly. I began to use a cane for stability, and I leaned on it more heavily as the term wore on. One day in October, I parked my car across the street from a store where I planned to make a purchase, and then found that I couldn't walk the short distance. I was reaching a crisis point.

Due to the peculiarities of the tumor's growth, my left side was, and still is, weaker than the right. By the fall of 1976, my left hand and arm began to weaken; fortunately, I'm right-handed. My other symptoms increased in number and severity. Muscle stiffness and weakness had by then affected most of my torso, and I noticed for the first time that the weakening of my chest muscles had affected my breathing. Talking for long periods of time tired me, and I had to make extra efforts when lecturing. Despite the physical problems, I maintained my normal teaching schedule, for I drove to work and parked in a garage that adjoined my building at Columbia. I had lots of bodily infirmities, but I was not really disabled—yet.

In mid-October 1976, I went to my neurologist for a checkup, and he was disturbed at the deterioration in my condition. The time had come, he told me, when surgery was necessary regardless of the dangers. The radiotherapy had killed off some of the tumor, but not all of it, and it had produced negative side effects. The radiation apparently had caused edema, or swelling, of the tumor, which increased the pressure on the spinal cord. And to aggravate the condition further, it had raised watery cysts on the tumor. Immediate steps were needed to reduce the pressures on the cord, and he told me that it was imperative that I enter the hospital as soon as possible. I arranged for colleagues to cover my classes for the remaining six weeks of the semester, and reentered the Neurological Institute at the end of October. Although I was admitted on an emergency basis, I had to wait two weeks for an opening on the surgery schedule. During this time, I had nothing to do except wonder why I couldn't have been waiting at home and school. But hospitals don't work that way, so I sat and chafed. I was still outwardly optimistic and

41

had some hope that the operation would do what radiation had not done. I also thought that I might be out in time to finish the last week or two of the semester. Neither rosy wish came true.

After ten days in the hospital, I finally learned that I would have surgery in a few days. The surgeon and the anesthesiologist came to my room the day before with permission forms, and I received the usual preoperative preparations. At 6:30 the next morning, I was sent, properly sedated, to the operating room. The routine is familiar to anybody who has undergone surgery, and it needs no recounting. My next memory after being told to breathe in deeply was a brief awakening in the operating room. I returned to full consciousness the next morning in the intensive care unit. Yolanda was there, waiting for me to awaken, but I fell asleep again after only a few minutes. Later that day, I was wheeled, bed and all, back to my room on the fifth floor, where I would recuperate for the next five weeks, long after the fall 1976 semester ended.

My mood during the first four days after the operation was ebullient. It was as if I had come back to life—I was born again. I was able to raise my left foot several inches off the bed, and I even walked, supported by a nurse, across the hall to the bathroom. My left leg was still weak, but I was able to lift it off the floor and move it ahead of the right one; it didn't just drag. If I could do all these things only a few days after surgery, I figured, then there was no limit to what I would be able to do in a few weeks.

My feelings of euphoria went beyond simple optimism. The shell of protection that I had built around my emotions melted, and my defenses—a wall built of humor, acerbity, and cynicism—fell before a welling of emotion that was a total departure from my usual state of mind. I seemed able to reach out and touch, almost embrace and engulf, everybody around me. I loved people, all people! This was not a matter of self-delusion, for others commented on it. It is difficult to recall those feelings now, but it seemed that the sharp edges of my self had become porous and weak. People could reach into me more easily, and they, in turn, were more vulnerable to me. I did not set hard

42

borders around my identity; I was suffused with a kind of peace-fulness, almost a sense of joy. It was all very strange.

It was evident to me even at the time that I was going through a kind of religious experience, the sort of conversion phenome-non that sends sinners to the altar rail, their faces streaked with tears, to state loudly that they have found Jesus and are giving their lives to Him. I had been raised as a Democrat and a Roman Catholic in that order of importance, for ours was not a very religious family, and the closest I had come to this kind of feel-ing was during my boyhood, immediately after going to confes-sion. But I remembered also that I had gone through a similar experience, though in a minor key, during the Columbia student strike of 1968, and others had reported the same mood as part of the Woodstock "happening" of 1969. Curiously, it was in my recollection of the events of spring 1968 that I found an under-standing of my postoperative euphoria.

The student uprising occurred eighteen years ago, and it was totally misunderstood as an antiwar protest by the media, which promoted and prolonged the strike by nightly newscast hyper-bole. Today, it is poorly remembered by those who went through it, and it is ancient history to a new student generation bound for the M.B.A. degree. There is little doubt that the Viet-nam War accentuated the 1968 unrest in the United States, but it is important to remember that, in the same year, the police killed a number of student demonstrators in Mexico City, and French student and worker strikes paralyzed parts of Paris. In response to those conservatives who saw a sinister Red influence at work, it is worth recalling that the French Communist party was outraged by the renegades. As for those on the Left who saw the protests as reactions to capitalist oppression, it is impor-tant to reflect that this was also the year that Soviet tanks en-tered Czechoslovakia to crush the Prague Spring.

If the Vietnam War was not the root of the discontent of the 1960s, what, then, was? I suspect that the least common denominator of all the movements was rebellion against the proliferation of an increasingly intrusive and impenetrable bureaucracy, which in our modern age has enveloped business and education as well as government. Bureaucracy renders the world of authority opaque, remote, and incomprehensible. It imposes a structure and an order that is impersonal and inimical. And it defies reduction to the far simpler notions about society that most people carry around in their heads. To make matters worse, bureaucracies are devised as rational, systematized organizations, but in operation they often are as capricious, arbitrary, and irresponsible as Wonderland's Red Queen. It is significant, therefore, that in the Berkeley Free Speech Movement of 1964, the forerunner of the later demonstrations, the students were voicing opposition to the *megaversity*, University of California President Clark Kerr's unfortunate term for the big, modern, multiservice university. Some students complained that they were being treated as "things," their identities and educations reduced to punch-card status. They found that their acts of defiance could stop bureaucracy in its tracks, and, however briefly, usher in a period of anti-order.

This, indeed, is what happened at Columbia in 1968. The student strike was less an antiwar protest than one of the first forays against the emerging postindustrial order, an attack by social Luddites on a remorseless future, a society as impersonal and dehumanized as an anthill. The strike forced the suspension of the formal university structure, and the society of students and faculty moved into a mode that the anthropologist Victor Turner called *communitas*.[1] In *communitas*, formal rules governing social discourse are suspended and people relate to each other affectively and diffusely—which beneath the jargon means that they no longer hide behind narrowly defined and formal rules of conduct, but rather meet as whole and caring people. This quality of social relations drew the Columbia faculty, which is notoriously compartmentalized and self-absorbed, into a true community for the first and only time, and it produced an unprecedented solidarity between students and some of their professors, at least in my department. In several others, however, 1968 caused rancor and deep division. It was this sense of immediacy and closeness to others, this lapse of the physical and symbolic barriers between people, that gave the

Columbia campus a kind of pre-Woodstock quality in 1968. Despite police raids, misguidedly intransigent administrators, and endless, awful student speeches, that special quality makes the year pleasant to remember. And it was this quality of oneness with others and inner wholeness that returned to me with intensity immediately after my surgery.

Communitas, in Turner's usage, is a phase in the process of ritual—it is a religious phenomenon. But there is no such thing as a purely "religious" attitude or behavior. Rather, what we call religion is a collection of ordinary human acts and dispositions to which certain sentiments and meanings have been attached. And the set of attitudes and feelings embodied in *communitas* bespeaks the negation of what I discussed in the last chapter as a rage for orderliness: There is also a need to suspend temporarily these mundane orders that we accept as natural and to move into a state of reversed time and sentiment, a mode of antistructure, free from the dead weight of authority and convention. It gives an exhilarating feeling.

All social systems are to some degree repressive, and mechanisms exist in most societies for a suspension of rules and a transformation of identities. This can be done through religious ritual, but it can also be achieved imperfectly through alcohol and drugs. Or it can happen as part of a return to life from a difficult phase, as with people recovering from serious illness or trauma. The sense of renewal that I, and countless others, have felt while recovering from surgery is part of a very normal aspect of psychological and social processes. I had been born again, returned to the world a new person. The decomposition of my identity by illness was healed. I was whole again in body and mind. It is a bright delusion, a shift into a kind of counterworld, flight to which makes the common-sense world bearable. The common-sense world, however, is the one in which we work and breed, and the one to which we must always return. And I returned from my own surgery-induced state of *communitas* to the "real" world after only five days, when I underwent a severe relapse.

Despite my high spirits and heady optimism after surgery, the operation had only the limited aim of decompressing the upper part of the spinal cord, thus protecting my hands and respiration. The surgeon had performed a laminectomy on the top half of my spine, leaving the bottom part of the tumor for a possible future procedure. He then removed a certain amount of dead tumor, but he was unable to remove all or even most of it. The most important part of the operation was a widening of the canal itself, relieving the pressure on the cord and also giving the tumor room in which to grow. It was this relief that I had felt immediately after the operation, but the respite proved to be short.

On the fifth day after the operation, I awakened with a deep sense of fatigue. One of my first impressions was that the strength of the muscles around my chest had decreased, affecting my breathing and the strength of my voice. I could no longer raise my left foot as far as I had done the day before, and my upbeat mood had turned into depression. My private-duty nurse immediately called the surgeon and the neurologist, and, as they examined me, their facial expressions confirmed that the operation had gone sour. What had happened was never made clear to me by my doctors, although one opined that it must have been due to something *I* did. In all likelihood there had been postsurgical edema, which, perhaps only temporarily, had created new pressures on the cord. But pressure for mere minutes can destroy nerve cells and cause irreversible damage, and I would never regain the high ground of those first four days.

The operation's immediate effect, then, was to leave me in worse condition than before, although a longer-term assessment would indicate that the dilation of the spinal canal would protect me for a while from future growth. The condition of my arms and legs had deteriorated, and I was no longer able to walk, even if supported by a cane or crutches. I could negotiate short distances with a walker, but otherwise my movements were limited to a wheelchair. This was not as big a setback as it may appear, for I was probably only a few months away from this stage before the surgery. The spasms in my legs and torso were also more intense, and I had a great deal of sensitivity, tingling, and pain in my arms and hands, especially on the left side. For the first time, I was experiencing true disability.

To explain the source of my problems, a short review of the landscape of the cortex is in order, starting at the bottom of the spinal cord and working upward. (My tumor, as I have noted, extended from the second cervical vertebra [C2] to the eighth thoracic one [T8], or from the top of the neck to the middle of the chest.) A spinal cord lesion (any damage, whether inflicted by disease or accident) at the level of the four sacral vertebrae near the base of the spine can cause malfunction of the bowel and bladder sphincters. This can result in incontinence or, more commonly, inability to eliminate wastes. Laxatives and catheterization of the bladder may be required. In cases of complete failure of the bladder sphincter, catheterization two or three times a day may be necessary, although this courts the risk of introducing a bladder infection. As an alternative, the sphincter can be removed, allowing the bladder to drain continually. Men who have had a sphincterotomy are able to put a condomlike sheath on the penis, which is attached by a hose to a plastic leg bag. For obvious anatomical reasons, women cannot do this, and they often suffer continual dampness.

Above the sacral vertebrae, a lesion at the level of the five lumbar vertebrae affects the legs, producing weakness or paralysis and necessitating a cane, crutches, or even a wheelchair. How much malfunction ensues depends on how badly and in what way the cord is damaged.

Next upward are the twelve thoracic vertebrae. A lesion from T6 to T12 causes weakness of the abdominal muscles and growing insensitivity to touch below the waist. Such a lesion also affects all functions below that level, because the sacral and lumbar areas are also cut off from regulation by the brain. Thus, a thoracic or cervical lesion usually affects bowel and bladder function and causes paraplegia. Damage at the level of T2 to T5 renders the chest and torso muscles flaccid and can substantially reduce lung capacity. The patient retains full use of the arms and hands, however, allowing considerable mobility and the prospect of independent living, albeit in a wheelchair.

Spinal cord damage in the cervical area results in quadriplegia, its severity contingent on the kind of lesion and its level. The hands and fingers are weakened and stiffened by damage from T1 to C6 and C7, and splints often are needed to keep the fingers from curling inward. A lesion at the C5 to C7 level wastes the arms as well as the hands, and this may prevent the patient from moving independently from wheelchair to bed or toilet, thus necessitating the help of family members or paid aides. Finally, a lesion at the level of C4 weakens the diaphragm, causing difficulty in speaking. Damage at C3 may destroy breathing, making a respirator necessary. This has been a sketchy description, for spinal cord damage affects every aspect of body function, but I will add other relevant details as my narrative progresses.

I said in the first pages of this book that every case is different, and my own does not fit neatly into the above scheme. My spinal cord was compressed, not severed, and in 1976, the tumor was either pressing lightly—or not at all—on the upper part of the cord. Even though the tumor extended to C2, my breathing was still good, showing that there was no heavy compression at that level. I also retained a good deal of bowel and bladder function, although I had to have my bladder catheterized on a few occasions. I was paraplegic, but I could stand and walk a little, and my arms and hands, especially on the right side, were still dextrous and reasonably strong. In fact, so many capabilities remained that I was considered a prime candidate for physical therapy.

After a week of postoperative recuperation, I commenced rehabilitation. It is ironic that neurosurgery and clinical neurology are among the most prestigious of medical specialties, whereas physiatry, or rehabilitation medicine, ranks among the lowest. It has been speculated that the poor prognoses of most neurological patients is what makes this one of the less glamorous areas of medicine, an idea seemingly confirmed by several medical students and interns, who told me that the field was too discouraging to attract them. They wanted to heal people, and there were not many spectacular cures in the neurological area.

But this attitude could apply with equal, or even greater, force to neurology and neurosurgery. Oliver Sacks, himself an eminent neurologist, says that neurologists can only examine and diagnose: "It is essentially a passive science." [2] Indeed, it could be argued that rehabilitation medicine has had an excellent record of restoring function to neurological patients beyond the predictions and hopes of these primary-care people. Stroke patients have been restored from immobility to normalcy, and victims of spinal cord injury have gone from helplessness to independent living. But there are no miracle cures in rehabilitation. If a patient improves, it is by degrees so fine that they are imperceptible from day to day. There are no magic potions—just hard, agonizing work.

The Neurological Institute houses a small physical therapy unit consisting of a sixteen-bed rehabilitation floor and a "gym" with a number of large leather mats raised on platforms some twenty inches above the floor. Two patients can be treated on each mat. There are also a number of weight-lifting stations, parallel bars used as handrails for people practicing walking, a leg-exercising machine, and an assortment of other devices. An adjoining room contains the occupational therapy group. Physical therapists and occupational therapists are both state-licensed professionals whose tasks differ mainly in the parts of the body they treat. Physical therapists are concerned with the rehabilitation of the major muscles of the legs, trunk, and arms, while occupational therapists concentrate on the hands and, to a certain extent, the arms. The O.T. room had a few tables as well as shelves full of the gadgets, tools, and supplies with which the patients worked. I underwent both occupational and physical therapy every weekday, but I did not live on that floor. My room, a small semiprivate one with a single bed, was located on another floor, and my nurse had to wheel me back and forth. My floor was more lonely than the rehabilitation floor, but, as I was to find out later, it was far more conducive to recovery.

My first session of physical therapy lasted twenty minutes. I was brought to one of the leather mats in a wheelchair and told to transfer myself from the chair to the mat. Even with the arm of the wheelchair removed, the gap of about six inches between chair and mat might as well have been the Grand Canyon. I was afraid to traverse it. The therapist told me to place my right hand on the mat and the left on the seat of the wheelchair, then push down, raising my body and shifting my weight to the right, moving my backside onto the mat. I tried it a couple of times, but I had no confidence that my left arm would be able to sustain my weight and push me over toward the mat. Just at the point of making the critical move over the six-inch-wide abyss, my nerve would fail. The therapist realized that I couldn't do it, so he helped me onto the mat, saying that we would try again tomorrow.

As I lay flat on my back on the mat, the therapist held one of my legs by the foot and knee, pushed my thigh against my trunk, and told me to push back until my leg was extended again. He exerted counterpressure against each of my pushes, and after only about five of them, I had to rest. We then worked on the other leg for a few minutes, at the end of which I was as exhausted as if I had just run a mile. The therapist was ready to continue, but I wasn't. We didn't try for a transfer from the mat back to the chair, but I was able to get to my feet with the aid of a walker and then slowly back around and into the wheelchair. Physically drained and psychologically daunted by the long, hard road I faced on the way to even a limited recovery, I returned to my bed a half hour after I had left it.

As the days passed, my therapy sessions lengthened, and I was able to transfer from chair to mat ever more frequently, although there still were occasions when my self-confidence failed. We began to work on the arms, pushing against the therapist's counterpush and lifting wrist weights. I also commenced exercises in which I would painfully struggle to my hands and knees and then rock my body back and forth. The leg exercises became at once longer and less exhausting. Even though I pushed back harder than before, the therapist seemed chronically dissatisfied with my performance—urging, cajoling, and nagging me onward until I would just give up and stop. I soon discovered that this was standard procedure, part of a game played by therapist, patient, and, often, an audience. One day, for example, a young paraplegic woman was helped to her feet, given a walker, and told to walk. After about five steps, she told the therapist, who was walking just behind her, that she was tired and wanted to stop. The therapist told her that she was giving up too soon and ordered her to continue. The other therapists and their patients echoed him, telling her that she could do it, forming a cheering section as she struggled onward. She soon stopped again, this time begging to be put back in the wheelchair, but the therapist was adamant. Finally, after she broke down in tears and showed signs of collapsing, the chair was brought up behind her and she fell into it. Everybody in the gym applauded, and she wiped away the tears and grinned in triumph.

The patients soon get into the spirit of the game, knowing that today's painful overreach may become tomorrow's routine accomplishment. Some even see it as a way of defying their doctors, who tend toward quite conservative prognoses. A couple of people told me, with obvious relish, as they hobbled along with the aid of two canes, that their doctors had told them that they would never walk again. Others proudly moved their arms and hands in refutation of a neurologist's prediction of total paralysis below the neck. The doctors, of course, are delighted about such "errors," for the pessimism of their predictions is often calculated carefully to prevent overoptimism, and perhaps crushing disappointment, on the part of the patient. There is some danger that doctors' predictions can also induce despair, and the exhortations of the therapists are meant to counteract the sometimes overpowering urge to give up.

Rehabilitation differs from other branches of medicine in the degree to which the patient is involved in his own treatment. Ideally, he is active, not passive, and he must try continually to outdo himself. To a degree, the patient is responsible for his own recovery, and this has many positive aspects. The negative side of patient responsibility, however, is that if his efforts can yield improvement, then any failure to improve can be an indication that he isn't trying hard enough, that he is to blame for his own condition. This load of culpability is often added to a lingering suspicion among family and friends that the patient was responsible, somehow or other, for what happened to him. And the patient, too, is often beset with guilt over his plight—a seemingly illogical, but very common, by-product of disability. In this way the patient's inner circle can escape a sense of remorse over his travails, and the medical establishment can absolve itself of responsibility for the failure of its procedures. Scapegoating, of course, is by no means limited to the handicapped. "He (or she) brought it on him(her)self" is a time-honored sentiment at funerals (along with, "I never speak ill of the dead, but . . ."), conveying the message that if you avoid doing what the deceased did (such as jogging or not jogging, as the case may be), then you won't die—at least for a while. It is a technique by which people assure themselves that the same unhappy fate will not befall them—a form of whistling in the dark.

This technique of *Blaming the Victim* is also the title of a biting sociological polemic written in 1972 by William Ryan.[3] Ryan documents the ways in which the poor are blamed for their own misery in the United States. Certain anthropologists and sociologists have written of a self-perpetuating "culture of poverty," providing scholarly justification for the belief that people receiving public assistance are responsible for their own misery and are beyond help. This attitude protects the American ideology that determination and hard work bring success. The converse of this notion is that those who do not succeed just did not try; *they* are to blame for their problems, not the rich or the economic system. Thus, we are urged to believe that the prevalence of the female-led family among black Americans is the source of the high unemployment and crime rates among their young men. The causality should be reversed: Technological change since 1960 has resulted in a dramatic decline in unskilled and semiskilled jobs, and prejudice and poor education keep blacks out of the skilled fields. This isn't laziness—it's economic genocide.

The parallels between the scapegoating of the poor and that of the disabled are obvious. Curiously, the similarity carries over into education. Right-wing critics of Head Start have long maintained that, given the social problems of the black family, such programs would do little good. Echoing this, in 1985 a federal political appointee claimed that the mainstream education of handicapped children was "immoral," because it diverted scarce resources away from pupils who could use them better. Pursuing this strange logic, she went on to say that the life circumstances of the disabled were appropriate to their inner nature. In short, they deserved their lot! We never did find out whether this was supposed to be due to predestination or bad karma, for she was driven from office by an angry Congress.

The impulse to yield to apathy and surrender to disability has sources other than guilt and gloomy prognoses. Most rehabilitation patients are, to some degree, depressed. These are generally nonclinical conditions—there are no neuroses—for the average neurological patient in rehabilitation has good reason to be depressed. Anybody can empathize with the deep depression of the twenty-one-year-old who often shared a mat with me, and who could barely bring himself to speak about his condition. He had been a perfectly healthy young man, with a good job as a lineman for the telephone company, when he and a friend walked into a New York City bar after work one evening. They were drinking beer peacefully when a gunman walked in and announced a holdup. The two friends sat still, but somebody disturbed the robbery, a wild shot was fired, and my fellow patient's spinal cord was nicked by the bullet. In a second's time, and without warning or reason, he was transformed from happy youth to chronic, lifelong cripple. I hope that he found some cheer in life since that time, for he surely foresaw none then. But he wasn't mentally ill; quite the contrary. He would have had to have been deranged to have felt happy.

Among people who have undergone sudden traumatic damage to the spinal cord, a period of "mourning" is quite common.[4] This phase is precipitated by a profound sense of loss and is accompanied by the onset of melancholia. Most shake off the depression and after several weeks or months start to readjust, however painfully, to their new circumstances. My own condition came on so slowly that I never went through mourning;

those with multiple sclerosis, which also has a gradual onset, similarly seem exempt from mourning, although they have terrible problems adjusting to the initial diagnosis. On the other hand, many of the patients in the rehabilitation unit had suffered strokes, which are not only sudden but beget a distinctive psychological state, the centerpiece of which is depression. They, too, have understandable reasons for feeling depressed, but they reach far lower depths than the other patients and for longer periods, sometimes entire lifetimes. Even without seeing the telltale sagging of the muscles on one side of the face, it is easy to spot them, for they sit in their wheelchairs with woebegone, vacant expressions, staring off into the distance at nothing—or perhaps nothingness. They were the most intractable of people in rehabilitation, and the therapists often had to shout just to get through the haze of funk that seemed to surround them visibly. To complicate matters, stroke sufferers have a much more common tendency to deny their illness than do other neurological patients.[5] This ranges in intensity from unwillingness to talk about it, to failure to accept what has happened, to outright denial that anything at all is wrong. Some stroke patients go so far as to claim that they are normal and capable of standing up and walking. They just don't want to. Needless to say, these are the worst prospects for rehabilitation. But whatever the ailment, and whatever the degree of denial, therapists must breach imposing psychological barriers to reach their patients and enlist their cooperation in the long, tedious process of reconstructing their bodies.

In contrast to the stroke and trauma patients, my mood was more positive and upbeat, despite the setback. I was making visible progress in therapy, and I was able to walk with a walker from my room to the nurses' station, a distance of about fifty feet. I was also doing well in occupational therapy, although I thought some of the exercises ridiculous. Nonetheless, visitors to our house still scrape their feet on the doormat that I made in O.T. Yolanda is the only person who knows its origins, a sign of the care I have taken to keep secret the indignities visited upon me in my disability—a privacy that ends with the publication of this book, and one reason why writing it has been so difficult.

Neurosurgery is of a different order from most other invasions of the body. People who have had open-heart surgery are usually out of the hospital in two or three weeks, and one man I know went home a month after he received a heart transplant. I stayed for six weeks after surgery, and probably would have been kept longer were it not for the approach of Christmas. That is considered a short stay, for more deeply damaged individuals might spend six months to a year in one of the rehabilitation facilities scattered throughout the country, where they rebuild capability and learn their ADLs (Activities of Daily Living). These are periods of restoring body strength and of reeducation in the basic skills of living. Depending on the type and gravity of the disability, the patient learns how to get on and off the toilet, how to dress, how to shave, how to brush his teeth, how to get in and out of bed, how to carry out all the minor chores of life with minimal dependence on others. These are activities that most of us—or, I should say, most of you—carry out unthinkingly, automatically, but each presents a great challenge to a paraplegic and almost an impossibility to a quadriplegic. I will have much more to say about this in the chapters that follow, for I only felt the full impact of my disability as the tumor grew and my musculature atrophied.

Beyond the need for postoperative physiotherapy, the aftereffects of any surgery on the central cortex can be profound. All surgery is an assault upon the body, aggravated by anesthesia hangover, but neurosurgery hits at the very seat of body regulation and coordination. Recovery was slow and painful, and I was often afflicted with a deep sense of fatigue and ill-being— the upshot of the surgery, six hours under anesthesia, and daily trips to the gym. Visitors and reading took up much of my free time, but I also spent hours lying in bed, my mind turned inward in contemplation of what had happened to me. It was then that I first pondered the question of whether my metamorphosis had any meaning. It was the beginning of this book.

As the time approached for leaving the hospital, I had very mixed feelings. I was, of course, delighted to be going home, but I would be returning to familiar surroundings in a different body. Would this new body survive in the old environment? Would I be able to adapt? And Yolanda? Would I not be a great burden to her, and would she be able to handle it? There was nothing at all in my rehabilitation that prepared me for the psychological and social challenges I would face, but this is a major failing of most rehabilitation programs in this country. Moreover, my physical rehabilitation also failed to retrain me for the future, because, at that time, I retained most of my upper body function and a good deal of my lower. I had no idea how I would adjust in the event of further deterioration, for I hoped there would be none. This vain wish got me into considerable trouble four years later.

The rehabilitation may have been shortsighted, but it was adequate for my neurological condition, at the time. It looked as if I could go home with ease, although I knew that our comfortable old house would now be an obstacle course. Just to begin with, somebody would have to carry me up the steps and into the house. It was with such thoughts that I left the hospital on December 22, just three days before Christmas of 1976.

3

THE RETURN

For there is one belief, one faith,
That is man's glory, his triumph, his immortality—
And that is his belief in life.
                    —Thomas Wolfe,
                      "This is Man"

It was only after my return to the real world of familiar settings, objects, and people that the dimensions of my disability were brought home to me. The hospital had been another universe, a place inhabited by sick people, many in far poorer condition than myself, and by their custodians. It was an alien place where our *significant others*—a jargony but apt term for those closest to us whose expectations most strongly shape our actions—shifted subtly from family, friends, and associates to doctors, nurses, and aides. Visitors to those long hospitalized often find that the patients are rather indifferent to news of the outside world, preferring instead to talk about other patients in the room, their doctors, and the events of the floor. Visitors and patients find themselves speaking to disinterested listeners, and the conversations have two subject matters, two universes of discourse. This happens in many encounters, of course, but when it occurs between people who share a common past, it disturbs the relationship, making the visit a trying experience. Moreover, it introduces an undercurrent of estrangement and contributes to the separatist quality of the sick role. Visits tend to become minor rituals, the settings of an interplay between alienation and agonized attempts to maintain solidarity.

Beyond its social system, the hospital presented an artificial physical environment, consisting in my case of bed, wheelchair, and the physical therapy room. It was a simplified setting in which there were no steps or other obstacles to mobility and few tasks to accomplish. I went to physical therapy every morning, carried out my exercises, returned to the room, practiced walking with my walker, got in and out of bed, and ate. I shaved myself, but most mornings I was wheeled to the shower room in a special chair and washed by a nurse. None of this was very demanding. The hospital was a physical and social cocoon in which rehabilitation was directed toward helping me regain certain minimal bodily functions that I had not yet lost completely, but leaving me quite unprepared for a return to a world that I last knew as somebody with only a bad limp.

The day of my release, December 22, 1976, was bitterly cold, and after two months in the overheated hospital, the wind seemed to cut through my clothing. My son Bob drove the car up to the hospital entrance, my wheelchair was parked and braked next to the front passenger seat, and I was faced with my first challenge in the outside world: how to get into a car. I grabbed the open door and pulled myself into a standing position, then made a quarter turn to the right and sat down heavily on the seat. I pulled my left leg into the car with my hands and swung my right one in after it on its own power. In the process, my body became unbalanced, and I began to fall to the left, a problem that recurred every time we turned a corner. I soon learned to compensate for the weakness of my torso muscles by holding on to the armrest and leaning against the turns. It occurred to me then that if I had to learn how to be a passenger all over again, how would I be able to relearn driving?

Our home in Leonia, New Jersey, is only a few miles from the Columbia-Presbyterian Medical Center and not far from the George Washington Bridge. It is a short trip geographically but a long voyage culturally. Yolanda and I are both New Yorkers, and initially we had wanted to rent an apartment on Morningside Heights, near Columbia, but when we moved East in 1963, Columbia's housing office never even acknowledged my application. So we gave up and moved to the suburbs. At least New Jersey was nearby, and we were home from the hospital in ten minutes. Bob parked in the driveway and I got out by reversing the entry procedure. Our next-door neighbor came out to greet me, and he and Bob carried me and the wheelchair up the seven steps to the front door. It was clear that I had to find a better way of getting in and out, or the house would become my prison.

Ecological science tells us that our physical surroundings are not inert and immutable entities, but sets of living, shifting relations between people and objects. And every last bit of our architecture, every avenue of public access, from sidewalk to subway, has been designed for people with working legs. Our house had not changed in any way, nor had we modified it to accommodate my new limitations. But my body had changed radically, altering completely the ecology of the house and household. Ours is an old two-and-a-half-story dwelling. Before my hospitalization, Bob had the attic room, our daughter, Pamela, slept in one of the three second-floor bedrooms, Yolanda and I occupied another, and we used the third as a study. There are also two full baths on the second floor, with the only tubs and showers in the house. But all these things were located at the top of a staircase that had suddenly become unscalable.

The ground floor has the kitchen, dining and living rooms, and a south-facing room with four windows, familiar to all northerners as the "sun room." This was used as a TV and guest room. It kept the television set carefully segregated from the rest of the house, and a sofa-bed folded out for overnight company. Shortly after moving into the house, we converted the room's large storage closet into a half bathroom containing a toilet and a sink. Since the first floor now provided all the basic facilities, and since I could not climb stairs, it was decided that I would sleep in the guest room, at least until we could make alterations.

The alterations were never made. We thought briefly of installing an elevator, but the cost would have been prohibitive. It would have made better economic sense to move to a one-story house. At the time I left the hospital, I could have used a stairway lift, although I would have had to keep one wheelchair at the bottom and another at the top. I was afraid, however, that if my condition deteriorated, I wouldn't be able to get on and off the lift's little chair. This came to pass only three years later, so we would have thrown away several thousand dollars. All discussions about getting me up to the second floor ended, however, when Bob and one of his friends carried me upstairs one day in 1977, and I discovered that the bathrooms would have to be torn up to make them accessible. That was the last time I saw the second floor of our house.

Due to the peculiarities of the tumor's growth, the left side of my body has been much weaker than the right. At that time, my right leg was capable of rising from one staircase step to another and had enough strength to straighten out and pull my body up. But the paralysis of the left leg had advanced so far that it couldn't follow. No matter how hard I tried, I couldn't raise that leg to the next step, thus ending my stair-climbing days. Much the same thing happened when I walked. My right leg and foot could move forward in a fairly normal step, but the left foot dragged on the floor. Not only was I unable to lift the left leg off the floor and swing it ahead of the right, but the weakened foot drooped downward, and I was unable to raise my left toes. For this reason, my walking consisted of stepping out with my right leg and dragging the left almost, but not quite, even with it. The hospital's orthopedic shop fitted me for a brace that was supposed to elevate my left foot. A simple one-piece device made of molded plastic with a Velcro strap, it cost $250 (all apparatus for the disabled is exorbitantly priced on the premise that insurance or public funding will pay the bill) and was an utter failure.

Nonetheless, I was able to use a walker to navigate the fifty-foot round trip across my living room. In order to allow me to drag my left foot, however, we had to take up the rugs. This also made it easier to push the wheelchair, for riding on carpets is much like pedaling a bicycle over a sand dune.

My upper body and arms were too weak for crutches, so I relied completely on a walker. The walker also allowed me to get to my feet to make transfers. I didn't like to sit in the wheelchair—first because they aren't very comfortable, and second because they are, quite appropriately, the symbol of disability—and I usually spent my days in a recliner chair. To make the switch, I would park the wheelchair near the recliner, set the brakes, pull the walker in front of me, rest my left hand on it for balance, and raise myself upright by using my legs while pushing downward on the wheelchair arm with my right hand. Getting up usually required all four limbs, although on good days I could manage with legs only. Most of the levering was done by the right side, with the left serving to provide balance. The most perilous part of the operation was the middle, at which point the arms had finished pushing and the legs had to take over completely. By the time I got past the midpoint, the legs would begin to exert their maximum strength. Once they were locked in a straight position, I could stand for quite long periods. After getting to my feet, I would take a couple of steps to the chair, swivel around in a counterclockwise direction, and sit down. A clockwise turn was impossible because it would have required me to lead off with my left leg, which was immovable. Due to this, I often had to swivel through 270 degrees when making a transfer.

The same techniques got me in and out of bed, and I had little trouble using the convertible sofa-bed, low though it was. Getting up in the morning was harder to manage, however, than lying down, for all the usual reasons and a few more. My room is small, and the position of the bed made it necessary to arise on the left side. To get up, I would lift my left leg with my hands, move it over the edge of the mattress, and drop it on the floor, swinging the right leg after it. I then had to push my body into a sitting position with my weak left arm, sometimes making several tries before succeeding.

The bathroom presented a different problem. I had no difficulty pushing my wheelchair up to the sink, but toilet seats are too low for paraplegics. This problem was easily solved, however, by installing a grab bar next to it and setting an elevator seat on top of the bowl.

I was quite self-sufficient back in that stage of my disability. I could dress my upper body, though I never did master pants and shoes, and I took care of most of my personal needs. I shaved, brushed my teeth, sponge-bathed most of my body, and used the toilet without assistance. Yolanda would go off to work for the day, leaving lunch in the refrigerator, and I would fend for myself. I even managed to reheat coffee. The only time that I required help was in getting dressed in the morning and undressed at night.

My reentry into the real physical world went quite smoothly. My reemergence into society was another matter. During the winter-spring semester of 1977, I was on sabbatical leave from Columbia, a period during which I had planned to write, but I spent most of it in recuperation. I should note in passing that, during the first ten years of my illness from 1976 to 1985, and through the course of several hospitalizations, I never received a single day's sick pay. I always had the rare bad luck to go into the hospital during leaves—until spring of 1986, when I had to take sick leave. I had intended to start working on this book during my fall 1984 semester sabbatical, but I spent three months in the hospital instead. And so it was that, in the harsh and unrelenting winter of 1977, I sat snowbound in the house, surveying my wrecked plans and my equally wrecked body.

It was during these months that I began to think seriously about death, although not for the first time. The apprehension of mortality is a condition of our consciousness, a first premise of our sense of time, an axiom in our grasp of selfhood, a constant factor in all our plans, and a thread that is woven through the very fabric of our being. Death makes manifest life and all its values; rather than being only a negation of life, it creates it and makes new life possible. Without death, the very concept of life would be meaningless. Total life, or pure being, and death, or pure nothingness, are one and the same: This is what old Hegel meant when he said that the work of the world lies in Becoming. And we humans differ from other animals in our certain knowledge of our own mortality, an awareness that is driven home to us by both direct observation and verbal account. Perhaps it is that distinctive possession of humans, language, with all its rich imagery and its wonderful capacity for hyperbole and fantasy, that makes death so vivid a presence in our thoughts—or that allows it to lie in our unconscious minds swathed in symbols, ready to leak out in dreamwork or pervasive, unarticulated anxiety. We humans are fearful, even morbid, creatures, for our intelligence escalates forebodings into terror. This apprehension would render us immobile were it not for the mind's great gift for repression and cultural institutions that deny, displace, and disguise mortality. But I had become a student of these stratagems, not a believer, and I accepted as simple fact that death is not an existential state on a different plane, but plain nothing. And whereas in my younger days the concept of death had been an abstraction, at the age of fifty-three and in catastrophic illness, it had become an ever-present reality.

The knowledge of death, its fear and anticipation, does something else that is rarely perceived: It is one of the premises of human loneliness. Nothing is quite so isolating as the knowledge that when one hurts, nobody else feels the pain; that when one sickens, the malaise is a private affair; and that when one dies, the world continues with barely a ripple. One can die very much alone in America, for the end often comes in hospitals, where the patient is attended only by medical people and life-support machinery. As if in recognition of this, the dying show increasing disinterest in the world, and the world reciprocates by segregating them systematically. We are not alone in this; the Eskimo used to abandon the old and infirm, and in certain African groups, the terminally ill were killed if they took too long to die. This kind of euthanasia is still frowned upon in our society (in

63

1985, a Florida court convicted a seventy-five-year-old man of first-degree murder for killing his wife, who was in an advanced stage of Alzheimer's disease), so instead we isolate the dying with heavy sedation. It is now common in our society for death to come during a drug-induced semicoma. Society thus excludes and denies death by placing distance between the living and the moribund.

Death and aloneness make common cause, a sociological truth known since 1897, when Émile Durkheim, the French founder of modern sociology, published his book *Suicide*.[1] In this work, Durkheim demonstrated statistically that people who dwell in single-room isolation are more prone to commit suicide than are those who live in families; loneliness begets alienation, a breeding ground for self-destruction. But as we well know, suicide is also quite common among people who are integral parts of families, and often for reasons arising directly from familial ties. That this is not solely a product of civilization's ills was brought home to us by our Mundurucu research. The people were loath to talk about suicide, for they consider it a calamity, but in every case they related, our informants attributed it to estrangement from kinfolk. In such statements as "her sisters spoke badly of her," the Mundurucu expressed the belief that the most destructive forms of alienation involve our closest bonds.

My own meditations on death arose directly from the sense of separation and isolation that overwhelmed me after I entered the cloistered life of the wheelchair. I would gaze out of the hospital window at people hurrying along the streets as if they were a species apart. A young resident in a jogging outfit ran past, and I thought, *There but for the grace of God go I*, which summed up my theological views on the matter. I felt that I was marching—or rolling—to a different drummer. I had visitors constantly every day that I was in the hospital, and I was surrounded by family at home, but I nonetheless was invaded by a profound sense of removal. My family felt this acutely, and their concerns for me became tainted by hurt and bewilderment. I believe that this fundamental orientation toward society is not unique to me but is a fixture of the consciousness of all disabled persons, a subject we will come to later. What is important to consider at this juncture is the fact that my personal isolation conspired with the gravity of my health to conjure visions of escape through death.

Despite thoughts about suicide, I was not truly depressed. I ate and slept well and kept my sense of humor. I even made jokes about my physical condition, although the laughter of my audiences was usually thin, strained, and polite. Suicide seemed to me to be a practical affair, a pragmatic way out of a disastrous situation. There I was, at fifty-three years of age, faced with a future of deep and permanent disability; my best hope was that it would get no worse, but I knew, somewhere in the back of my mind, that this was an illusion. The reality was that the tumor was still alive and well, and that it would grow slowly and inexorably, grinding everything before it like a glacier. The final stages of this progressive paralysis would be a form of bare survival that I repress even now from my thoughts. There seemed to be only one way to escape the tumor before it reduced my body to total inertness.

Suicide is a simple matter for the able-bodied, but I soon realized that it is a very difficult feat for the motor-disabled. How would I go about buying a gun or poison without my wife knowing about it? And even if I were able to make the purchase, where would I hide it? And for how long would my fingers have the strength to pull the trigger? The irony of my situation was that by the time I would want most to end my life, I would be physically unable to do it. I began dreaming of suicide. In my most vivid dream, I pulled the pin from a hand grenade, held it to my chest, and counted; I awoke at the count of three. The death wish was real, but the dream stuff wasn't. I did wartime service in the Navy, and the only hand grenades I ever saw were in John Wayne movies. And the Duke, along with the latter-day warrior Sylvester Stallone, never got nearer to war than a movie studio. Where, then, did my dream end and grade-B films begin? Dreams, movies, and even real life are all built upon mirages.

Yolanda soon guessed what I had been thinking and confronted me with it. Her argument was direct and simple. Instead of telling me about all the good things of life, which would have been Pollyanna-ish, she told me what my death would do to her and our children. It was not my private affair, for, although our lives in theory may belong to us, they are mortgaged. My life belonged to my family, primarily, and also to many others. This realization that suicide is a crime against the living, against life itself, ended the matter. Our children were seventeen and eighteen years old at the time, and their educations were unfinished. My professor's salary supported us comfortably, but little was left over for savings. It became clear that I could not afford to die yet. Beyond economic necessity, the family and all the people close to me needed me alive more than ever, for, especially in my impaired state, I had to symbolize for them the value of life itself. It is this trust that suicide betrays, for it is the duty of all of us to die defiantly, with a touch of class.

I was badly damaged, yet just as alive as ever, and I had to make the best of it with my remaining capabilities. It then occurred to me that this is the universal human condition. We all have to muddle through life within our limitations, and while I had certain physical handicaps, I retained many strengths. My brain was the only part of the central cortex that still worked well, but that also is where I made my living. *Disability* is an amorphous and relativistic term. Some people are unable to do what I do because they lack the mental equipment, and in this sense, they are disabled and I am not. Everybody is disabled in one way or another. And even though my growing paralysis would one day end my active participation in the affairs of the world, I could still sit back and watch them unfold. Anthropology had made me into a voyeur of all things human, and it had educated me to their elusive beauty and transiency. Being alive was simply too interesting. I decided to rejoin the world.

My attitude toward life and living nonetheless had become altered during illness. My very existence was under such constant threat that I began to look on each day, week, month, and year as a gift. Since that time, I have lived in the present, each day a lifetime's work, each birthday a miracle. By the time the

end arrives, I have reasoned, my condition may have so deteriorated that death, too, would come as a gift. And so I no longer
feared death. It has since that time stayed with me as an old
friend, in uneasy companionship with my reaffirmation of life. It
was all very Irish.

There had been another time when I had been forced into a
similar decision. Children of alcoholics never seem to learn from
example, and there is a well-documented tendency toward re-
creating the sins of the fathers. There are those who believe that
this indicates that predisposition to alcoholism is hereditary and
genetic—a plausible idea, but like most such quick-and-easy explanations (racial intelligence, female subordination, for example), no one has located the gene. A more provable explanation
is that a penchant for drink is generated in the complex relationships of the nuclear family, through socialization within a ma-
trix of weak men and strong women whose capacity for
maternal love is exceeded only by their talent for suffering.

This was indeed the sociology of my family, and the drift into
a routine of heavy drinking was made easy, almost inevitable, by
transition into young manhood during wartime. Weekend liberties were devoted to womanizing and drinking, in that order of
importance, and in inverse order of success. Once aboard ships,
we drank only on the infrequent occasions of port visits, but we
then tried to make up for all the months at sea. In Manila we
drank "whiskey" that was so vile that you had to hold your nose
while swallowing it—a concoction that cost a dollar a shot and
could cause blindness, literally and figuratively. After World
War II, we returnees went on a prolonged liberty, although most
of us eventually lapsed into a more relaxed routine of frequent,
nonbinge drinking. It was the accepted medium of socialization,
a pattern that has continued to the present day.

My seagoing days had not, however, ended. After leaving my
ship in Pusan, Korea, for a trip on a merchant vessel to San
Francisco and home, I swore that I would never again set foot
on a ship. Mustered out, I signed up for my GI Bill unemployment insurance (called "the 52-20 Club" because we were eligible for a year's benefits at twenty dollars a week) and waited for

Columbia College (also paid for by the GI Bill) to start in September 1946. And so it was that one day in early May 1946, I went to the U.S. Employment Service to get my regular check, was asked the ritual weekly question about my occupation, replied again that I was a maritime electronics technician—smugly confident that I would never be placed—and was told by the clerk that they had a job for me. Four days later, I was standing on the deck of the *General Goethals*, a troopship in the United States Army's Military Sea Transportation Service, as she backed into the Narrows from a Staten Island dock, bound for Bremerhaven, Germany.

Despite recurrent dreams thereafter of being shanghaied, it was a pleasant four months, for we sailed as civilians, not as military. The pay was good, the work was light, and on each side of the Atlantic I led the sailor's life. On one trip to Germany, I stayed drunk for five days on champagne, bought with occupation scrip easily obtained through black market sales. There were no tomorrows and there were no places of abode. There were only movements and ports—and drinking. I went to sea again on an Army troopship in the summer of 1949 and entered Columbia's graduate school that September, my identity badly split between sailor and scholar.

By the end of the Second World War, I was a career drinker, but I had not yet fallen into alcoholism. This stalked me during the 1950s while I was teaching at Berkeley and captured me in Africa during a field trip in 1959–60. There I assumed the white man's burden of daily drinking, a chore that did not impede my research too much and that was so much a part of the colonial ambience that I was able to delude myself about what was happening to me. Yolanda—whom I married in 1950, in my pre-alcoholism days—saw it but didn't know what to do about it, hoping I would settle down when we returned to California. She was to learn that alcoholism is not a situational thing, a response to some transient stress or concern. The true alcoholic does not drink to ease problems or anguish; rather, he carefully, though often unconsciously, creates such dilemmas in order to justify drinking. Alcoholism is a disorder that is imbedded in the character structure, an intricate process that has a predetermined trajectory, a compulsion that leads teleologically toward oblivion.

The serenity of academic life has been badly overstated, and the Berkeley Department of Anthropology in and about 1960 was riven by conflicts of personality and ideology, each sometimes masquerading as the other. As a result, there was a mass exodus of younger faculty members to the East, a return migration that I joined in 1963, when I went back to Columbia as a full professor, nine years after receiving a doctoral degree there. Back at Columbia, I found old friends, a supportive department, and academic tranquillity. I had attributed much of my increased drinking at Berkeley to the atmosphere of the department, a gambit that had the virtue of allowing me to blame others for my ills. But almost as if to refute deliberately such neat displacement of responsibility, I began to drink even more heavily in New York. Yolanda became more alarmed, our arguments grew more frequent and intense, and I sank deeper into alcoholic depression.

There are many, many alcoholic academics, for one can get away with it for years as long as he meets his classes in reasonable sobriety. As part of an attempt to deny my alcoholism and to conceal it from others, I took care never to miss classes and always to start the day's drinking in the late afternoon. My writing had virtually halted, but I was able to coast for some time on the publications I had amassed in order to get tenure at Berkeley. The list was longer than most because a few senior professors there wished me to leave, and I had to publish profusely to avoid perishing at a tenure review. I was thus able to maintain a facade of good teacher and productive scholar, bon vivant, convivial drinking companion, and comic-in-residence. Instead, however, I was turning into a solitary drinker whose own drinking buddies were unaware that they had lost him.

I knew I couldn't keep up the act much longer, for it was draining me psychologically. Everybody maintains an identity that is in part counterfeit, and we all devote considerable time and energy to perpetuating the fraud and selling it to others. But when the private reality begins to parody the public image, and when the two are radically discordant, a break must come. The scholar and the sailor in me were on a full collision course, and the sailor was winning. By 1966, I found myself on the verge of destroying both my marriage and my career, and of drifting downward and inward into the limbo in which my father had dwelled.

The alcoholic lives in an isolation chamber, a closed world that finds its organization and rallying point in the satisfaction of a craving. He schedules his time and budgets his resources around the need to drink, and he must conceal these machinations from others. In the process, he becomes a schemer and a dissembler, an artificer of masquerades, the practitioner of a secret life. Of graver consequence, he manipulates others as unwitting accomplices, and even enlists protesting spouses as star performers in a family drama. Their very opposition to drinking becomes part of a struggle that serves to incite more drinking; the family reorganizes itself around the alcoholic, but the alcoholic arranges his existence around himself. It is a journey into aloneness, a fall into one's self, a return to primal narcissism. It is a touch of death in the midst of life.

Christmas of 1966 had provided an appropriate setting and excuse for my annual holiday drunk, but it left me more depressed than ever. By the fall of that year, my drinking was beginning to get out of hand. I was losing the middle-class alcoholic's art of maintaining a "plateau" of intoxication in public and was getting too drunk at the wrong times and places. It frightened me, for I knew all the symptoms. I had not yet "bottomed out," as they say in Alcoholics Anonymous, meaning that I had not lost job and family, nor was I ever found in a gutter. But I had indeed started down a steep slide in that direction. I was a person without a future, tottering at the edge of an abyss, poised at a crossroads; I knew that a decision had to be made and that there could be no turning back. I was thinking these thoughts on the night of December 27, with a quart bottle of beer in front of me, when I stood up, poured the beer down

the sink, and went to bed. I have not had an alcoholic drink of
any kind since that time.

I had taken an unusual approach in quitting liquor. I didn't go
to a psychiatrist or internist beforehand or afterward, I didn't
take tranquilizers, and I didn't go to Alcoholics Anonymous. I
stopped cold turkey, and I never had a relapse. I deliberately
went it alone for very specific reasons. The semireligious quality
of the AA experience repelled me, and I didn't want to join the
more secular Columbia chapter because I suspected that the
confessions of my fellow professors would sound like scholarly
papers, sententious and self-serving. Besides, I had watched AA
friends become absolutely dependent on the organization. When
they fell off the wagon, it was a gesture of defiance against the
group, and their returns were acts of submission and contrition.
Implicit in this is a kind of parent-and-rebellious-child relationship, and the sometimes-overt message: "I am weak; the group
is strong." Instead of orienting their lives around a bottle, they
reorganized them around AA, and in so doing perpetuated their
estrangement from the rest of society.

I did not want simply to form a new dependency, to lose myself again. Rather, I wanted to refind myself, to relocate and
redefine the lost sense of who and what I was. The years of
drinking had sapped my will and destroyed my self-esteem; I
thought of myself as weak and spineless, irresolute and selfish—
and I was right. And the years of camouflage had left me uncertain of my very identity. Through my solitary route to sobriety,
supported only by Yolanda, I had done much more than dry
out. I had reasserted mastery over my actions, reestablished my
feelings of self-worth, and redirected my perspectives to the outside world. The months following my quitting were difficult, but
I will not recount that tale, for it no more belongs in this book
than do the details of my tippling. Of relevant interest, instead,
is alcoholism as an existential state, and in this respect, the pain
of withdrawal was a necessary price of redemption, a validation
of a hard-won freedom. It was a manumission from myself that I
had forged myself. I had found a part of me that I never knew

71

existed. And as I was to do again ten years later, I had refound, then rejoined the world.

To this day, I cannot say why I fell into alcoholism, for I have never undertaken psychoanalysis, and I know better than to try to analyze myself. There is a substantial body of analytic opinion, however, that sees male alcoholism as a means of maintaining and reasserting masculinity, and of denying tendencies toward passivity and identification with the mother. Paradoxically, this protest is expressed through a profound oral dependency. The syndrome is thus in large part the product of growing up with a weak, even despised, paternal role model and a strong, deeply ambivalent maternal figure. My own history was a minor variation of this major theme.

That my drinking could be a result of masculine overprotest has given me a comforting link with the men of many primitive societies, including the Amazonian Mundurucu, who believe that they keep their dominance over women only through their jealously guarded control over certain sacred musical instruments. These instruments, according to myth, once were owned by the women, who used them to dominate the men; but in the world's first sexual revolution, the men seized them from the women and reversed the roles of the sexes. The myth and its rituals are a parable of the dominant mother and the struggles of the male to escape her and protect an ever-fragile masculinity. Bullroarers and sacred flutes—all wrought in wonderfully phallic shapes—are aspects of a universal masculine attempt to maintain the myth of their superiority, along with men-only clubs, overbearing husbands, and drunken sailors. We are all brothers under the skin; it would be funny but for the costs to all of us, male and female alike. What unites the world's men in brotherhood, then, is the fragility of male identity. We are forced to lose the mother as both love object and model, and our grasp of masculinity is threatened continually by the urge to fall back into her folds, to reverse our hard-won autonomy and relapse into dependent passivity. The mother figure is both the giver of life and its destroyer, and the strivings of men for dominance are expressive of this fear and weakness.

Alcoholism is one way out of the male dilemma. It is an asser-
tion of camaraderie, a vehicle for the expression of male solidarity, a scenario for masculine bravado. My own identity was
split between the fake rollicking sailor on eternal shore leave
and the inner frightened thirteen-year-old kid whose mother had
left him. But unlike our primitive brethren, who express the dilemma of maternal loss and threat in ritual fashion, the way of
the alcoholic is antisocial. It may start in conviviality, but it ends
in brooding isolation. The alcoholic resolves the problems of the
world by resignation from it; he asserts his masculinity by a
parody of dominance; he drowns his inner fears in forgetfulness;
he palliates the delicacy of social relations with atrophy and sev-
erance; he turns away from life and retreats into death. It is
suicide, however bloodless and gutless, just as much as the suicide I contemplated while recovering in the winter of 1977. But
this, we will see, is not the only similarity of alcoholism to
disability.

Despite all the problems of being a department chairman during Columbia's troubles of the late 1960s and early 1970s, I
never lapsed from my dry regimen. Nor was I even vaguely
tempted to do so during the long course of my illness. I knew
that alcohol creates the very depression that the drinker tries to
erase, and I knew that if I began serious drinking, I would fall
into self-pity, a state of mind that is fatal to the disabled. One of
the things that sustained me was the memory of my father, unemployed throughout the Depression, sitting at the dining room
table playing his endless games of solitaire—bereft of friends,
work, and even family. My own alcoholism reconciled me with
him some twenty years after he died, and his memory was an
element in my deliverance—the first of my two resurrections.

As the snows melted in the spring of 1977, I came back slowly to
life. My friends in Leonia rallied around me, and I never lacked
visitors in the first few months. As time wore on and the perma-
nence of my disability became explicit, however, I saw less of
them. Students also came out to see me, and we even held a few
oral exams in the living room. But there were many days when I

wished they would all go away and leave me to lick my wounds quietly, setting up an oscillation between involvement and withdrawal that has bedeviled me all my life, but one that grew stronger in my illness. During this time, I stayed in the house, a prisoner of the weather and the front and back steps, doing little except nondirected reading: a mix of anthropological books and journals, doctoral dissertations, science fiction, mysteries, and anything else that came my way. I had no projects or plans other than the sick person's duty of getting better.

My first venture outside came in March, when I received the Mark Van Doren Award for teaching from the students of Columbia College, my alma mater. The award came at a critical time in my recovery, for it went directly to the most vulnerable part of my ego. The main teaching vehicle of the university professor is the lecture, a performance that is an agony for some and a delight for others, particularly hams. I belong to the latter category. A dull fifty-minute lecture can be an affliction for audience and lecturer alike, as I well knew from having sat through, and delivered, many. On my off days, I have looked out at my hapless victims—sitting in a halo of numb boredom, slackjawed and staring, fighting to stay awake—and despised myself for inflicting this on fellow humans. Even more disheartening has been watching the pre-meds and pre-laws voraciously absorbing and recording every last miserable, unadorned, and uninterpreted "fact," happy that at last there was some "meat" in the course, something that could be spewed back on an exam, some bedrock of certainty on their march toward a fat practice. I have tried instead to preserve the romance of the discipline, to use American culture as a primary source of data, and, above all, to keep the students awake. That I may have succeeded to some degree was enormously gratifying, and the award dinner was the high point of my teaching career. It was an elixir for the wounded self.

The dinner was also my reentry into society, and it was appropriate that it happened in connection with a ritual. More than an academic honor, and however unintended, it functioned as a *rite of passage*, Arnold van Gennep's term for those ceremonial occasions marking a person's transition from one social identity to another.[2] Rituals dramatize the change from boy to man, from girl to woman, from single to wedded, from living to dead—and mine saw my reemergence in public as a disabled person. As is always appropriate for such events, there was a gathering at which everybody ate and drank at a communal feast, and the participants were the people with whom I interacted as a professor. Present were the president of the university, the dean of the college, all of Columbia's anthropologists, and a large number of graduate and undergraduate students. My wife and children were there, feeling a bit left out, but no other relatives. The only neighbors present were also members of the Columbia community. The turnout was large and the audience response even warmer than seemed called for; I soon realized that I was being hailed more as a survivor than as a teacher—it was a homecoming rite. No matter, the sentiment was genuine, and I again felt myself to be a part of the university. The ritual congregation did much to soothe my profound uneasiness at once again meeting my old associates, though in a different body. But I was to find out that it had by no means purged me of all the ritual novice's pollution of identity.

With the advent of warmer spring weather, I had a ramp built from the back door, giving me access to the car. I was fully capable at that time of driving with hand brakes and throttle, but I never did try it. I had enough strength in my hands to use the controls, and power steering made turning easy, but my head and shoulders could only swivel ninety degrees, and I would have had to use the mirrors for backing. Besides, I believed that my motor reflexes had become slow and erratic, making me a potential menace to everybody on the road. To complicate matters, I did not have enough upper body and arm strength to pull the wheelchair into the car after me, so I would need help at each end of a trip. I decided to let Yolanda do the driving. It was a bad decision, one guided by fear and withdrawal rather than by realistic appraisal of the alternatives, but by the time I realized this, driving had become impossible. I am not certain how much practical transportation I would have gotten from my own driving, but it would have yielded huge psychological benefits in a returned feeling of mastery and power. I had forgotten how much my sense of being in control was heightened by driving during my limping days. Instead, I now had taken a step backward.

The inability to drive was more than a retreat from mobility, for it was one more step away from spontaneity and the free exercise of will. Whereas I could once act on whim and fancy, I now had to exercise planning and foresight. This was true of even the simplest actions. In transferring from wheelchair to toilet, bed, or armchair, I had to park in a carefully chosen position and set the brakes, lest the wheelchair fly out from under me. I next would plot my strategy for getting up, choosing my supports with care, calculating the number of steps it would take to reach my target. When I no longer was able to walk or stand, such transfers became impossible, and my movements became even more restricted, the obstacles more imposing. The house had become a battlefield, my movements well-thought-out strategies against a constant foe.

My outside activities were even more constrained. I could go nowhere without a driver, usually Yolanda or Bob, and every trip entailed logistics. If we wanted to eat at a restaurant or go to the movies, we had to call first to make sure there were no steps. And if we wanted to stay at a motel, we had to measure the width of the bathroom door, for my wheelchair is twenty-six inches wide. Whenever we traveled, we needed equipment. Aside from the wheelchair, we had to transport the walker, bedpan, urinal, and assorted accessories. As my condition worsened, the list grew longer. Gone were the days when we would give in to a sudden urge to go somewhere or do something. But gone also were the days when I could wander into the kitchen for a snack or outside for a breath of fresh air. This loss of spontaneity invaded my entire assessment of time. It rigidified my short-range perspectives and introduced a calculating quality into an existence that formerly had been pleasantly disordered.

My experience was by no means unique, for all motor-disabled people have the same complaint. In his brilliant book *Awakenings*, Oliver Sacks reports on one woman, brought out of a sleeping-sickness–induced coma by the drug L-Dopa, who had lost control of her body movements.[3] Once headed in a certain direction, she would keep on going until stopped by a wall or a piece of furniture. If she wanted to reach a certain chair, she would point herself toward it, but if she could not do this in a straight line, she had to bounce her body off a wall and into the chair, as in a billiards carom. Each move had to be planned carefully; each traverse was an exercise in ballistics. This is an extreme case, to be sure, but all para- and quadriplegics share her problems of planning, albeit to a lesser degree. As for entering the outside world, paralytics who don't own cars (and they are legion) have to arrange for transport days in advance. Some can be carried only in wheelchair vans, which are very expensive if hired and are paid for by Medicare only if used for medical appointments. They are hemmed in further by the schedules of their aides and the groups providing social and medical services. In all this, they are passive recipients, waiting for the world to come to them—in its own time, if at all.

By the beginning of May, I could no longer pose as an invalid in convalescence, and I was becoming bored and increasingly irritated at my wasted sabbatical. I began to think of writing something, anything. I have always been a two-finger typist—a crude method, to be sure, but I had pecked out several books with my two middle fingers. My left hand had, however, become too weak for our old manual machine, and I planned to buy an electric one. Hearing this, a group of Leonia friends banded together and bought me one. Like the computer I now have, it was a wonderful new toy, and I spent hours playing with it. Indeed, I thought, it would be fun to write a book. But about what?

In 1962, I had signed a contract with the Prentice-Hall publishing house to do a textbook on introductory general anthropology. My motivation at the time was simple: money. During the 1960s, anthropology course enrollments were soaring, and there were few textbooks. A widely used text could double or triple my income and relieve all my financial worries. By that time, my drinking was becoming increasingly costly, filling me with growing anxiety and insecurity about the future. Alcoholics are hag-ridden with fear and guilt—which they ease, of course, by drinking—and I had a bad case of the syndrome. Besides, being a child of the Depression, I never forgot what it felt like not to get quite enough to eat. A successful text seemed an ideal way out of my worries. When I sat down to write the book, however, I stared at the blank page in the typewriter and nothing happened.

Two things were wrong. Alcohol had eroded my ability to undertake a major writing project, and the book wasn't in me. A book on general anthropology must cover all of its four sub-fields—cultural anthropology, archaeology, physical anthropology, and linguistics—and I had lost interest in and current familiarity with all but the cultural branch. I had decided to stop teaching the other fields the day a student asked me to tell him the composition of DNA. "Protein," I answered. "What's protein?" he inquired. "Do you think I'm a chemist?" I responded. The very thought of having to go back to fossil man, arrow points, and syntax, as if I were starting graduate work all over again, dismayed me, blocking me from writing even the cultural sections. After four years, I had completed only the introductory chapter. It was at that time that I stopped drinking, and it might seem that I then would have been able to get on with the book, but the opposite resulted—I shelved the project. When the editor, an occasional drinking buddy, asked why, I told him, "Because I'm no longer afraid."

The unfulfilled commitment rankled me, however, and in the spring of 1977, while looking for a writing project, I hit on the idea of writing a text on cultural anthropology only, a subject I had been teaching for twenty-five years and that I knew so well that I could write a book about it off the top of my head. This had a real advantage, beyond being a solution to my search for a topic. My old research habits were out of step with my newly limited capabilities. I could no longer prowl the library stacks, I couldn't get to the books in my old second-floor study, and I was unable to reach those on higher shelves downstairs. This prevented me from continuing my rather chaotic, but quite effective, rummaging of past research. I needed a book I could write from memory, and this was it.

Starting in early May, I finished a 100,000-word manuscript at the end of August. Once embarked on the venture, I worked at it continually and feverishly. I do not tend to write calmly and methodically. Each work becomes a passion, and its writing more a physical necessity than a chore. Some of my books have had greater urgency than others. My *Dialectics of Social Life* [4] consumed me for eight months in 1968 and 1969. As it progressed, it seemed to take me over, until I felt as if the manuscript were writing itself through me. The book had its own rhythm and logic, and I just worked the typewriter. The 1979 work, to be titled *An Overture to Social Anthropology,* [5] possessed me also, but for different reasons. Whereas the subject matter and the turbulent period in which it was written accounted for the fervor of *Dialectics*, my personal circumstance made *Overture* an outpouring of a returned zest for life. I only realized this after I finished the book. Instead of the usual sense of quiet satisfaction, I became exultant, victorious. Amazed at my own reaction, I told Yolanda that I had never known that I had this much pride, or that it had been so threatened by my illness. "So what else is new?" she replied in true New York style.

To settle the old obligation, I sent the completed manuscript to Prentice-Hall through a friend who was at that time their anthropology consultant. He passed it on to the new editor (publishers change editors more often than their socks, much to the chagrin of authors) with a strong recommendation to publish it. But the editor didn't know about the old contract and was not interested at first in the book. He changed his mind when three outside readers reviewed it favorably. The book went into production, and the first copies appeared in 1978, only eighteen months after I started page one. It was successful enough to warrant a second edition in 1986, but it didn't earn the huge royalties I had dreamed of in 1962, for a flood of competitors had appeared in the intervening years, and nationwide anthropology enrollments had turned downward. But the book had served another purpose. It brought me back into the world of active scholarship and publishing. It served notice to my profession and my university that I was still alive, despite occasional rumors that I literally was quite dead. And it revitalized a shaken belief that I was still a person of worth, that I could still command the respect of my peers, that my life still had value, that some things remained unchanged.

I faced one more challenge that year: a return to teaching in September 1977. The class I had always enjoyed most was my introduction to cultural anthropology, which usually drew about 200 students and called forth all my bent for showmanship. I was much less concerned with what I would say than with how I would say it. My chest muscles had atrophied to such a degree that my ribcage was splayed. Talking for long periods tired me, and my voice had lost timbre and resonance, no longer projecting as well as it did. I was uncertain whether I could make it through a seventy-five-minute lecture, and I was apprehensive about the reaction of the students to a professor in a wheelchair. Would I be able to continue teaching?

On the first day of classes, my teaching assistant wheeled me to the lecture hall and up to the podium. The students looked at me in slight surprise, but after a few seconds resumed conversation with their neighbors. The moment of truth had come and gone with barely a murmur. Now came the hard part. Even with a lapel microphone, a healthy person has to raise his voice in a class of that size, and since this was the first day of the course, I had made plans in my own mind for an early bailout, some excuse for ending the lecture ahead of schedule in case I gave out. I called the class to order, introduced myself and the two teaching assistants, distributed syllabuses, described the course and its requirements, and then called for questions. This was my first escape point, but my voice was strong, and I was enjoying myself immensely; I kept going. As the lecture went on, I made a startling discovery: I was pushing the words out with my diaphragm, and my voice was gaining strength from the exercise. The students listened attentively, they laughed at the right points, and we all forgot that I was in a wheelchair. My sense of control and self-confidence surged back, and I talked for five extra minutes.

Academic people do three things: They teach, they do research, and they write. I was both teaching and writing in 1977, and I would undertake major new research in a few years. This again raises questions. Who is disabled? What is disability, anyway? When the 1980 census form arrived, I looked at the question that asked whether anybody in the household was fully disabled and checked "no." The question seemed to be one related to income rather than health, and I remained fully employed. My physical deficiencies would keep me from further research in the boondocks, but I was getting a bit old for that anyhow. Otherwise, I was neither "handicapped" nor "disabled" in my profession. And I took inordinate satisfaction in this.

There was more to my writing efforts than "publish or perish," for I had been given tenure at Berkeley and it was a condition of my Columbia appointment. Rather, it was an illustration of what I have called Murphy's First Law of Academic Careers, of which there are two phases: In the first, young academics are anxious over whether they will be discovered; in the second, the established ones are worried about whether they will be found out. This is what makes Sammy, and me, run, but my own struggles against decline were made more intense by my attempt to deny my disability. My overreach beyond the limits of my body was a way of telling the academic world that I was still alive and doing what I always did. And all my feverish activities in both academia and my community were shouts to the world: "Hey, it's the same old me inside this body!" These were ways for protecting the identity, for preserving that inner sense of who one is that is the individual's anchor in a transient universe. Many authorities on rehabilitation would look on this as a sign that I had failed to "accept" my disability, which often translates as unwillingness to become a good, passive client for their services. On the contrary, I knew what was wrong with me, and I knew that it would get worse, not better. I acknowledged the physical condition, but I would never accept its social limits or tolerate the eclipse of my past—nor should any disabled person do so, for it is the very basis of his or her struggle for life.

PART II
BODY, SELF, AND SOCIETY

4

THE DAMAGED SELF

As Gregor Samsa awoke one morning from uneasy dreams he found himself transformed in his bed into a gigantic insect. He was lying on his hard, as if it were armor-plated, back and when he lifted his head a little he could see his domelike brown belly divided into stiff arched segments. . . . What has happened to me? he thought. It was no dream.
                    —Franz Kafka,
                    The Metamorphosis

From the time my tumor was first diagnosed through my entry into wheelchair life, I had an increasing apprehension that I had lost much more than the full use of my legs. I had also lost a part of my self. It was not just that people acted differently toward me, which they did, but rather that I felt differently toward myself. I had changed in my own mind, in my self-image, and in the basic conditions of my existence. It left me feeling alone and isolated, despite strong support from family and friends; moreover, it was a change for the worse, a diminution of everything I used to be. This was particularly frightening for somebody who had clawed his way up from poverty to a position of respect. I had become a person of substance, and that substance was oozing away. It threatened everything that Yolanda and I had put together over the years. In middle age, the ground beneath me had convulsed. And I had no idea why and how this had happened.

I cannot remember ever before thinking about physical disability, except as something that happened to other, less fortunate, people. It certainly had no relevance to me. A disabled person could enter my field of vision, but my mind would fail to register him—a kind of selective blindness quite common among people of our culture. During a year that I spent in the Sahel and Sudan zones of Nigeria and Niger, a region of endemic leprosy and missing hands, feet, and noses, the plight of those people was as alien to me as were their language, culture, and circumstances. Because of this gulf, I had no empathy for them and just enough sympathy to drop coins into cups extended from the ends of stumps. A few pennies were all that it took to buy the dubious grace of almsgiving. It was a bargain, a gesture that did not assert my oneness with them, but rather my separation from them.

With the onset of my own impairment, I became almost morbidly sensitive to the social position and treatment of the disabled, and I began to notice nuances of behavior that would have gone over my head in times past. One of my earliest observations was that social relations between the disabled and the able-bodied are tense, awkward, and problematic. This is something that every handicapped person knows, but it surprised me at the time. For example, when I was in the hospital, a young woman visitor entered my room with a look of total consternation on her face. She exclaimed that she had just seen an awful sight, a girl who was missing half of her skull. I knew the girl as a very sweet, but quite retarded, teenaged patient who used to drop in on me a few times a day; we always had the same conversation. I asked my guest why the sight bothered her so much, but she couldn't tell me. She in turn asked why it didn't trouble me. Afer a moment's thought, I replied that I was one of "them," a notion that she rejected vehemently. But why did my visitor, a poised and intelligent person, react in this way? It aroused my curiosity.

There is something quite significant in this small encounter, for it had elements of what Erving Goffman called "one of the primal scenes of sociology." [1] Borrowing the Freudian metaphor of the primal scene (the child's traumatic witnessing of the mother and father in sexual intercourse), Goffman used the phrase to mean any social confrontation of people in which there is some great flaw, such as when one of the parties has no nose. This robs the encounter of firm cultural guidelines, traumatizing it and leaving the people involved wholly uncertain about what to expect from each other. It has the potential for social calamity.

The intensely problematic character of relations between those with damaged bodies and the more-or-less unmarked cannot be shrugged off simply as a result of the latter's ineptitude, bias, stupidity, and so forth, although they do play a part. Even the best-intentioned able-bodied people have difficulty anticipating the reactions of the disabled, for interpretations are warped by the impairment. To complicate matters, the disabled also enter the social arena with a skewed perspective. Not only are their bodies altered, but their ways of thinking about themselves and about the persons and objects of the external world have become profoundly transformed. They have experienced a revolution of consciousness. They have undergone a metamorphosis.

Nobody has ever asked me what it is like to be a paraplegic—and now a quadriplegic—for this would violate all the rules of middle-class etiquette. A few have asked me what caused my condition, and, after hearing the answer, have looked as though they wished they hadn't. After all, tumors can happen to anybody—even to them. Polite manners may protect us from most such intrusions, but it is remarkable that physicians seldom ask either. They like "hard facts" obtainable through modern technology or old-fashioned jabbing with a pin and asking whether you feel it. These tests supposedly provide good, "objective" measures of neurological damage, but, like sociological questionnaires, they reduce experience to neat distinctions of black or white and ignore the broad range of ideation and emotion that always accompanies disability. The full subjective states of the patient are of little concern in the medical model of disability, which holds that the problem arises wholly from some anatomic or physiological disorder and is correctible by standard modes of therapy—drugs, surgery, radiation, or whatever. What goes on inside the patient's head is another department, and if there are signs of serious psychological malaise, he is packed off to the proper specialist.

The medical people have had little curiosity about what I think about my condition, although they do know its sensory symptoms: a constant tingling in my hands, forearms, and feet and a steady, low threshold of pain in my legs. The discomfort is similar to the soreness and burning sensation experienced with torn or severely strained muscles and has much to do with the fact that the musculature of my trunk and legs is always in spasm, despite a generous daily dosage of muscle relaxants. The curious thing about this condition is that a knife could be run through my leg now and I wouldn't feel it. Aside from the tingle and the ache, my legs are otherwise numb and bereft of sensation. For the past four years, I have not tried to move them, for the tumor has long since passed the point at which therapy could maintain function; they don't even twitch anymore.

For a while, I occasionally tried to will the legs to move, but each futile attempt was psychologically devastating, leaving me feeling broken and helpless. I soon stopped trying. The average nondisabled person could be driven to the edge of breakdown if his legs were pinioned and rendered totally immobile for long periods, and an accident victim in a body cast finds his only comfort in the fact that his situation is temporary. I was saved from this, however, because the slow process of paralysis of my limbs was paralleled by a progressive atrophy of the need and impulse for physical activity. I was losing the will to move.

My upper body functions have suffered some impairment ever since the tumor became symptomatic, although the lower body has deteriorated at a much faster rate. In addition to the diminished lung capacity and, consequently, more rapid and shallow breathing, the musculature of the arms and hands has progressively weakened, their range of movement has become steadily more narrow, my fingers have stiffened, and my hands have become increasingly numb and insensitive to touch and temperature. In common with other paralytics, I have to be careful with hot water or hot dishes and pots, for I can burn myself without knowing it. To complicate the hand problems, the fingers of quadriplegics curl inward toward the palms, a process that by the spring of 1986 has made my left hand almost useless.

Beyond these physical symptoms, I have been overtaken by a profound and deepening sense of tiredness—a total, draining weariness that I must resist every waking minute. It starts in the morning when I struggle to awaken, fighting my way out of the comfort and forgetfulness of sleep into self-awareness and renewed disability. Facing the world every day is an ordeal for everybody, and it is no accident that strokes and heart attacks peak at 8 to 9 A.M. It is much worse to confront the day with a serious deficit. The wish to turn my back to the world continues through the daily ablutions, which grow longer and more tedious every year; it now takes me a quarter hour to shave. I am fully able to face life by 10 A.M., but by 4 or 5 P.M., I start to flag. Between these hours, I teach, talk to students, and attend meetings, after which I go home and lie down for a couple of hours. I also conserve strength by spending two, or at most three, days a week at the university. Like most professors, I don't pass the other days in idleness. I read student exams, reports, and doctoral dissertations; I keep up on books and journals in my field; and I do research and writing. But the professorial life allows me to work at home at my own tempo, sometimes while lying in bed. In no other line of work, I tell my graduate students, could such a wreck be 100 percent employed—it has to be an easy job!

But there is another aspect of my fatigue that cannot be eased by rest. This is a sense of tiredness and ennui with practically everything and everybody, a desire to withdraw from the world, to crawl into a hole and pull the lid over my head. The average person will recognize this wish, for everybody at some time or other feels that things have become too much to handle and he or she wishes for surcease, for even temporary remission. How tempting to tell all and sundry—family, work, and society—to go to hell and leave him alone. Who hasn't said this, even if only under his breath? When an ordinary citizen is overcome by these feelings every day and all day, however, his family and friends will urge him to seek professional help, for these are the sure symptoms of depression. In contrast, the deeply impaired harbor these urges chronically, sometimes because they are depressed but more often because they must each day face an inimical world, using the limited resources of a damaged body.

Many give in to the impulse to withdraw, retreating into a little universe sustained by monthly Social Security disability checks, a life circumscribed by the four walls of an apartment and linked to outside society by a television set. Constantina Safilios-Rothschild, a sociologist, has noted that disability may provide a pretext for withdrawal from work for some older workers dissatisfied and weary with their jobs, a kind of "secondary gain" bought at great price.[2] But this is not the source of the isolation; it's just making the best of a bad thing. Many other disabled people go forth to battle the world every day, but even they must wage a constant rear-guard action against the backward pull. This is a powerful centripetal force, for it is commonly exacerbated by an altered sense of selfhood, one that has been savaged by the partial destruction of the body. Disability is not simply a physical affair for us; it is our ontology, a condition of our being in the world.

Of all the psychological syndromes associated with disability, the most pervasive, and the most destructive, is a radical loss of self-esteem. This sense of damage to the self, the acquisition of what Erving Goffman called a "stigma," or a "spoiled identity,"[3] grew upon me during my first months in a wheelchair, and it hit me hardest when I returned to the university in the fall of 1977. By then, I could no longer hold on to the myth that I was using a wheelchair during convalescence. I had to face the unpalatable fact that I was wedded permanently to it; it had become an indispensable extension of my body. Strangely, I also felt this as a major blow to my pride.

The damage to my ego showed most painfully in an odd and wholly irrational sense of embarrassment and lowered self-worth when I was with people on my social periphery. Most of my colleagues in the anthropology department were old friends, some even from our undergraduate years, and they generally were warm and supportive. But people from other departments and the administration were another matter. During my first semester back at the university, I attended a few lunch meetings at the Faculty Club, but I began to notice that these were strained occasions. People whom I knew did not look my way. And persons with whom I had a nodding acquaintance did not nod; they, too, were busily looking off in another direction. Others gave my wheelchair a wide berth, as if it were surrounded by a penumbra of contamination. These were not happy encounters.

My social isolation became acute during stand-up gatherings, such as receptions and cocktail parties. I discovered that I was now three-and-a-half feet tall, and most social interaction was taking place two feet above me. When speaking to a standing person, I have to crane my neck back and look upward, a position that stretches my larynx and further weakens my diminished vocal strength. Conversation in such settings has become an effort. Moreover, it was commonplace that I would be virtually ignored in a crowd for long periods, broken by short bursts of patronization. There was no escape from these intermittent attentions, for it is very difficult to maneuver a wheelchair through a crowd. My low stature and relative immobility thus made me the defenseless recipient of overtures, rather than their instigator. This is a common plaint of the motor-disabled: They have limited choice in socializing and often must wait for the others to come to them. As a consequence, I now attend only small, sit-down gatherings.

Not having yet read the literature on the sociology of disability, I did not immediately recognize the pattern of avoidance. Perhaps this was for the best, as my initial hurt and puzzlement ultimately led me to research the subject. In the meantime, I stopped going to the Faculty Club and curtailed my contacts with the university-at-large. This is not hard to do at Columbia, as each department lies within a Maginot Line, everybody is very busy, and the general social atmosphere runs from tepid to cool. None of this is surprising, for it is also the dominant ethos of New York City. On the positive side, this same general mood allows one to work in peace. They leave you alone at Columbia, and I wanted more than ever to be left alone.

Withdrawal only compounds the disabled person's subjective feelings of damage and lowered worth, sentiments that become manifest as shame and guilt. I once suggested to a housebound elderly woman that she should use a walker for going outside. "I would never do that," she replied. "I'd be ashamed to be seen." "It's not your fault that you have arthritis," I argued. I added that I used a walker, and I wasn't ashamed—this was untrue, of course, and I knew it. But why should anyone feel shame about his disability? Even more mysterious, why should anyone feel a sense of guilt? In what way could I be responsible for my physical state? It could not be attributed to smoking or drinking, the favorite whipping boys of amateur diagnosticians, and it wasn't the result of an accident, with its possibilities for lifestyle culpability, the accusation that one bought it by living dangerously. No, I didn't do a damned thing to earn my tumor, nor was there any way that I could have prevented it. But such feelings are endemic among the disabled. One young woman, who had been born without lower limbs, told me that she had felt guilt for this since childhood, as had her parents (from whom she probably acquired the guilt). Indeed, a mutuality of guilt is the very life-stuff of the paralytic's family, just as it is, on a smaller scale, central to the cohesion—and turmoil—of all modern families.

Guilt and shame are not in fact as separate as they are often represented to be. In simple form, both are said to involve an assault on the ego: Guilt is the attack of the superego, or conscience, and shame arises from the opprobrium of others. Of the two, I believe that shame is the more potent. The sociologist George Herbert Mead wrote that an individual's concept of his or her self is a reflection, or, more accurately, a refraction, as in a fun-house mirror, of the way he or she is treated by others.[4] And if a person is treated with ridicule, contempt, or aversion, then his own ego is diminished, his dignity and humanity are called into question. Shaming is an especially potent means of social control in small-scale societies, where everybody is known and behavior is highly visible, but it is less effective in complex societies like our own, where we can compartmentalize our lives and exist in relative anonymity. But a wheelchair cannot be hidden; it is brutally visible. And to the extent that the wheelchair's occupant is treated with aversion, even disdain, his sense of worth suffers. Damage to the body, then, causes diminution of the self, which is further magnified by debasement by others.

Shame and guilt are one in that both lower self-esteem and undercut the facade of dignity we present to the world. Moreover, in our culture they tend to stimulate each other. The usual formula is that a wrongful act leads to a guilty conscience; if the guilt becomes publicly known, then shame must be added to the sequence, followed by punishment. There is then a causal chain that goes from wrongful act to guilt to shame to punishment. A fascinating aspect of disability is that it diametrically and completely reverses this progression, while preserving every step. The sequence of the person damaged in body goes from punishment (the impairment) to shame to guilt and, finally, to the crime. This is not a real crime but a self-delusion that lurks in our fears and fantasies, in the haunting, never-articulated question: What did I do to deserve this?

In this topsy-turvy world of reversed causality, the punishment—for this is how crippling is unconsciously apprehended—begets the crime. All of this happens despite the fact that the individual may be in no way to blame for his condition; real responsibility is irrelevant. This transmutation of body impairment into guilt is a neat inversion of the Freudian Oedipal drama. According to the psychoanalytic interpretation of the myth, Oedipus unknowingly kills his father and marries his own mother, for which crime the Fates pursue him to Colonnus, where he blinds himself. Blinding is seen as a symbolic form of castration, which is, in turn, the fitting punishment for incest.

According to Freud, in male socialization it is the threat of castration by the father—even if only a fantasized threat—that forces the child to relinquish and repress the guilt-ridden wish to possess the mother. It should be noted, however, that in the myth the father did not blind Oedipus; Oedipus did it to himself. What is usually forgotten in discussions of the Greek tragedy is that the father, after hearing from a soothsayer that his son would one day slay him, crippled young Oedipus. In fact, the name Oedipus can be translated as "Swollen Foot" or, loosely, "Gimpy."

That crippling can be just as proper a punishment for incest as blinding finds ethnographic support. In my own fieldwork, I recorded a Mundurucu myth in which a man who committed incest with his surrogate mother was physically deformed by her husband, a god, and later blinded. And among West African Moslems there is a widespread religious cult centered on a female succubus named Dogwa. (In Morocco, this cult figure is called Aisha Kandisha.) Dogwa is both nurturant and sexually seductive to her devotees; she is both mother and lover. In the former role, she can bring wealth to her followers, but in the latter, she jealously takes swift retaliation against infidelity. Appropriately, the punishment is crippling or blinding. It is worth noting that no father figure is involved here. Instead there is the ambivalent mother, the giver and nurturer of life and its potential destroyer. Incest, or even the unconscious wish for it, is a dangerous game, and I would hazard the guess that the unconscious, diffuse sense of guilt that so often bedevils the disabled arises in the first place from the chimerical notion that the crippling is a punishment for this repressed, elusive, and forbidden desire. There may be no such thing as Original Sin, but original guilt lurks in the dark recesses of the minds of all humans. These ashes of our first love are the basic stuff of the indefinable, unarticulated, and haunting sense that the visitation of paralysis is a form of atonement—a Draconian penance.

Paralytic disability constitutes emasculation of a more direct and total nature. For the male, the weakening and atrophy of the body threaten all the cultural values of masculinity: strength, activeness, speed, virility, stamina, and fortitude. Many disabled men, and women, try to compensate for their deficiencies by becoming involved in athletics. Paraplegics play wheelchair basketball, engage in racing, enter marathons, and do weight-lifting and many other active things. Those too old or too impaired for physical displays may instead show their competence by becoming "super-crips." Just as "super-moms" supposedly go off to work every morning, cook Cordon Bleu dinners at night, play with the kids, and then become red-hot lovers after the children are put to bed, the super-crip works harder than other people, travels extensively, goes to everything, and takes part in anything that comes along. This is how he shows the world that he is like everybody else, only better.

Becoming a super-crip, or super-mom, often depends less on the personal qualities of the individual than on very fortunate circumstances. In my own case, I was well established in my profession at the time of my disability, so my activity was just a matter of persistence. The real super-crips are those who do it all after they become impaired, like one woman who, after partial remission from totally paralytic multiple sclerosis, went on to finish college and then obtain a Ph.D. She refused to let the disease rob her of a future. There are many such people, but, like super-moms, they are still a minority. The vast majority, as we will see, are unable to conquer the formidable physical and social obstacles that confront them, and they live in the penumbra of society, condemned to lives as outsiders.

Afflictions of the spinal cord have a further devastating effect upon masculinity, aside from paralysis, for they commonly produce some degree of impotence or sexual malfunction. Depending on the extent of damage to the cord, the numbed genital area sends no signals to the brain, nor do the libidinal centers of the brain get messages through to the genitalia and the physiological processes that produce erections. This can result in total and permanent impotence, sporadic impotence, or difficulty in sustaining an erection until orgasm. There are some paraplegic men, on the other hand, who can maintain an erection but are unable to achieve orgasm, even after steady intercourse of a half hour to an hour. The effects of this on the male psyche are profound. We usually think of "castration anxiety" as an Oedipal thing, but there is a sort of symbolic castration in impotence that creates a kind of existential anxiety among all men. It is no accident that impotence is a major problem in those lands where masculine values are strongest, nor was it fortuitous that the new sexual freedom in America, with its emphasis on female gratification and male performance, has yielded a bumper crop of impotent men. After all, being a man does not mean just having a penis—it means having a sexually useful one. Anything less than that is indeed a kind of castration, although I am using this lurid Freudian term primarily as a metaphor for loss of both sexual and social power.

Most forms of paraplegia and quadriplegia cause male impotence and female inability to orgasm. But paralytic women need not be aroused or experience orgasmic pleasure to engage in genital sex, and many indulge regularly in intercourse and even bear children, although by Caesarean section. Human sexuality, Freud tells us, is polymorphously perverse, meaning that the entire body is erogenous, and the joys of sex varied. Paraplegic women claim to derive psychological gratification from the sex act itself, as well as from the stimulation of other parts of their bodies and the knowledge that they are still able to give pleasure to others. They may derive less physical gratification from sex than before becoming disabled, but they are still active participants. Males have far more circumscribed anatomical limits. Other than having a surgical implant that produces a simulated erection, the man can no longer engage in genital sex. He either becomes celibate or practices oral sex—or any of the many other variations in sexual expression devised by our innovative species. Whatever the alternative, his standing as a man has been compromised far more than has been the woman's status. He has been effectively emasculated.

Even in those cases in which the paraplegic male retains potency, his stance during the sex act changes. Most must lie still on their backs during intercourse, and it is the woman who must do the mounting and thrusting. In modern America, this is an acceptable alternative position, but in some cultures it would be considered a violation of male dominance: Men are on top in society and they should be on top in sex, and that's the end of the matter. And even in the relatively liberated United States of the 1980s, the male usually takes the more active role and the position on top. But the paraplegic male, whether engaging in genital or oral sex, always takes a passive role. Most paralytic men accept this limitation, for they discover that the wells of passion are in the brain, not between the legs, and that pleasure is possible even without orgasm. One man, who had enjoyed an intensely erotic relationship with his wife before an auto accident made him paraplegic and impotent, reported that they simply continued oral sex. The wife derives complete orgasmic satisfaction and the husband achieves deep psychological pleasure, which he describes as a "mental orgasm." The sex lives of most paralyzed men, however, remain symbolic of a more general passivity and dependency that touches every aspect of their existence and is the antithesis of the male values of direction, activity, initiative, and control.

The sexual problems of the disabled are aggravated by a widespread view that they are either malignantly sexual, like libidinous dwarfs, or, more commonly, completely asexual, an attribute frequently ascribed to the elderly as well. These erroneous notions, which I suspect arise from the sexual anxieties of their holders, fail to recognize that a large majority of disabled people have the same urges as the able-bodied, and are just as competent in expressing them. Spinal cord injuries raise special problems, but motor-disabled people with cerebral palsy, the aftereffects of polio, and many other conditions often can lead almost normal sex lives. That asexuality is also attributed sometimes to the blind underlines the utter irrationality of the belief.

Given the prevalence of such ignorance, I was pleased to read in 1985 that educational television was airing a film on sex among the disabled, and I made it a point to watch it. At the beginning of the film, there appeared on the screen a warning that there would be nude scenes, leading me to the happy expectation that a para- or quadriplegic would be shown making love.

Not so, for the only nudes were a couple of very healthy-looking young women. And, in deliberate counterpoint, most of the disabled people shown were grossly disfigured. It was a modern-day version of *Beauty and the Beast*, a film that served to perpetuate, not combat, a prejudice. The producers meant well, but they merely illustrated the depth of the problem. This episode reminded me that when one young woman began research among paraplegics, a female friend asked her, "But you wouldn't go to bed with one of them, would you?" These are indeed primal scenes.

One of Sigmund Freud's enduring contributions to our age was his rejection of classical philosophy's disembodiment of the mind. Instead, Freud started with a theory of instincts that located much of human motivation and thought in the needs of the body, especially the sexual drive. This was not a simple single-direction mechanical determinism, however, for Freud's theories held that causality is a two-way street. The human mind also uses its symbolic capacity to reach out and encompass the body, making it just as much a part of the mind as the mind is of the body. The body, particularly the more explicitly erogenous zones, becomes incorporated into human thought, into the very structure of the personality, and the sexual symbolism of pleasure and desire is used by the mind in molding one's orientation to the world. Sex thus invades thought, but is also intensified and transformed by thought. And so it is that the loss of the use of one's legs, or any other vital function, is an infringement also on the integrity of the mind, an assault on character, a vitiation of power.

The unity of mind and body is also an important element in phenomenological philosophy. This school, which arose in the early twentieth century from the writings of Edmund Husserl, sidesteps the old philosophical question of "how do we know the world [or reality, or truth]" and says that the world is whatever we make it out to be; it is created within the stream of conscious experience. And the way we experience and understand reality is in good part shaped by the language categories through which we sort out what we take to be real, and by the cultural symbolism through which we find significance and meaning in the mess of sense impressions continually bombarding us. Reality, then, isn't a hard-and-fast thing, the same for everybody, but a consensual matter, a social construct, that must be reaffirmed and re-created in all our interactions with other people. It would follow from this relativistic view of the human grasp of the world that people of different cultures inhabit somewhat different realities, as do people of the same culture but of radically different circumstances—people, for example, who can't walk.

In his 1962 book *The Phenomenology of Perception*, the French philosopher Maurice Merleau-Ponty states that the starting point for our apprehension and construction of the world is the body.[5] This goes beyond the obvious fact that our sense organs are parts of the body, for he stresses that the landscape of the body is, explicitly or implicitly, the means and the perspective by which we place ourselves in environments and experience their dimensions. As Simone de Beauvoir says, the body is not a thing, an entity separate from the mind and from the rest of the world in which it is situated. The body is also a set of relationships that link the outer world and the mind into a system. Merleau-Ponty illustrates this by reference to the phenomenon of the "phantom limb," the amputee's illusion that he still possesses the missing arm or leg. He writes, "What it is in us which refuses mutilation and disablement is an *I* [Merleau-Ponty's emphasis] committed to a certain physical and inter-human world, who continues to tend towards his world despite handicaps and amputations and who, to this extent, does not recognize them *de jure* [openly and avowedly]."[6] The amputee is missing more than a limb: He is also missing one of his conceptual links to the world, an anchor of his very existence.

Gelya Frank, an anthropologist, has written a life history of a woman born without her four limbs, documenting the laborious process by which she became "embodied" and grew to accept her condition and develop self-love.[7] Frank sees her as a kind of Venus de Milo whose beauty is curiously enhanced, like the statue's, by her lack of limbs. Embodiment is a problem for those born with deficiencies, but at least they can be socialized to their limitations from infancy. On the other hand, most paraplegics and quadriplegics come to their lot through "the slings and arrows of outrageous fortune" and have a different problem—they have to become reembodied to their impairments. And if the loss of function is grave enough, they may even have to become disembodied.

My own sense of disembodiment is somewhat akin to that of Christina, the "disembodied lady" discussed by Oliver Sacks in his book *The Man Who Mistook His Wife for a Hat*.[8] Because of an allergic reaction to an antibiotic drug, Christina lost all sense of her body—a failure of her faculty of proprioception, the delicate, subliminal feedback mechanism that tells the brain about the position, tension, and general feeling of the body and its parts. It is this "sixth sense" that allows for coordination of movement; without it, talking, walking, even standing, are virtually impossible. In similar fashion, I no longer know where my feet are, and without the low-level pain I still feel, I would hardly know I had legs. Indeed, one of the early symptoms of my malady was a tendency to lose my balance when I would take off my pants in the dark, something that happened to me often in my drinking days. Christina's troubles differ from those of the paralytic, however, for her loss has been more complete. Besides, she became disembodied while still capable of movement, and she compensated by using her eyes to coordinate her physical actions. Quadriplegics, too, must watch what they are doing, and I have spilled drinks held in my hand because my wrist had turned and my brain didn't register it. But by the time the paralytic's failure of proprioception is as complete as Christina's, the limb is no longer movable, and the condition is moot.

I have also become rather emotionally detached from my body, often referring to one of my limbs as *the* leg or *the* arm. People who help me on a regular basis have also fallen into this pattern ("I'll hold the arms and you grab the legs"), as if this depersonalization would compensate for what otherwise would be an intolerable violation of my personal space. The paralytic becomes accustomed to being lifted, rolled, pushed, pulled, and twisted, and he survives this treatment by putting emotional distance between himself and his body. Others join in this effort, and I well remember that after I came to from neurosurgery in 1976, there was a sign pinned to my sheet that read, DO NOT LIFT BY ARMS. I weakly suggested to the nurse that they print another sign saying THIS SIDE UP.

As my condition has deteriorated, I have come increasingly to look upon my body as a faulty life-support system, the only function of which is to sustain my head. It is all a bit like *Donovan's Brain*, an old science-fiction movie in which a quite nefarious brain is kept alive in a jar with mysterious wires and tubes attached to it. Murphy's brain is similarly sitting on a body that has no movement or tactile sense below the arms and shoulders, and that functions mainly to oxygenate the blood, receive nourishment, and eliminate wastes. In none of these capacities does it do a very good job. My solution to this dilemma is radical dissociation from the body, a kind of etherealization of identity. Perhaps one reason for my success in this adaptation is that I never did take much pride in my body. I am of medium height, rather scrawny, and militantly nonathletic. I was never much to look at, but that didn't bother me greatly. From boyhood onward, I cultivated my wits instead. It is a very different matter for an athletically inclined boy or a girl on the threshold of dating and courtship.

Those who have lost use of some parts of their bodies learn to cultivate the others. The blind develop acute sensitivity to sounds, and quadriplegics, who cannot handle heavy telephone directories, have a remarkable knack for remembering phone numbers. But of a more fundamental order, the quadriplegic's body can no longer speak a "silent language" in the expression of emotions or concepts too elusive for ordinary speech, for the delicate feedback loops between thought and movement have been broken. Proximity, gesture, and body-set have been muted, and the body's ability to articulate thought has been stilled. It is perhaps for this reason that writing has become almost an addiction for me, for in it thought and mind become a system, united in conjunction with the movements of my hands and the responses of the machine. Of even more profound impact on existential states, the thinking activity of the brain cannot be dissolved into motion, and the mind can no longer be lost in an internal dialogue with physical movement. This leaves one adrift in a lonely monologue, an inner soliloquy without rest or surcease, and often without subject matter. Consciousness is overtaken and devoured in contemplation, meditation, ratiocination, and reflection without end, relieved only by one's remaining movements, and sleep.

My thoughts and sense of being alive have been driven back into my brain, where I now reside. More than ever before, it is the base from which I reach out and grasp the world. Many paralytics say that they no longer feel attached to their bodies, which is another way of expressing the shattering of Merleau-Ponty's mind-body system. But it also has a few positive aspects. Just as an anthropologist gets a better perspective on his own culture through long and deep study of a radically different one, my extended sojourn in disability has given me, like it or not, a measure of estrangement far beyond the yield of any trip. I now stand somewhat apart from American culture, making me in many ways a stranger. And with this estrangement has come a greater urge to penetrate the veneer of cultural differences and reach an understanding of the underlying unity of all human experience.

My own disembodied thoughts are crude when compared with those of many people. A blind Milton painted sweeping landscapes of the heavens in *Paradise Lost*, and Beethoven crafted the Ninth Symphony despite—or perhaps because of— being deaf. And today one of the world's leading cosmologists, a Cambridge physicist named Stephen Hawking, travels through quarks and black holes in a journey across space and time to the birth of the universe. These are voyages of the mind, for Hawking has an advanced case of amyotrophic lateral sclerosis (familiarly known as Lou Gehrig's disease), which has left him with only slight movement in one hand and an inability to speak above a whisper. There are not many Miltons, Beethovens, and

Hawkings, however, and their example may be small comfort to a twenty-year-old quadriplegic who has made the mistake of diving into shallow water. For most disabled people, the loss of synchrony between mind and body has few compensations.

Many years ago, long before I became disabled, I was talking to a black anthropologist, a friend from our days as fellow graduate students, and the subject turned to race. In the course of our conversation, my friend said, "I always think of myself as being black, just as you always think of yourself as white." I protested this, saying that even though I did think of myself as white when talking to a black person, my skin color was not in the forefront of my conscious mind at other times. My friend didn't believe me. But I was neither mistaken nor misleading in my observation, for I grew up in and still lived in a white world. Whiteness was taken for granted; it was standard and part of the usual order of things. I lived in white neighborhoods; I sailed on a white warship (except for the officers' stewards, who were black); I went to white schools (P.S. 114 in Rockaway Beach never had a black student during my eight years there); and I work in a profession that is still ninety-five percent white. Why think of my whiteness when most of my contacts are with white people? The comedian Martin Mull once did a television program entitled "The History of White People in America," a howlingly funny title because its redundancy fractures logic. White is normal; it's what ethnolinguists call an "unmarked category," a word that is dominant within its class and against which other words of that class are contrasted. Why, I would no more have thought of myself as white than I would have thought of myself as walking on two legs.

Before my disability, I was a standard White, Anglo-Saxon, Agnostic Male (WASAM?), a member of the dominant part of the society. My roots in tattered-lace-curtain Irish Catholicism made me uneasy in academia, but I never gave much thought to the other components of my identity. My black friend was forced by the reality of white society always to think of himself as black. It was his first line of defense against a hostile environment. His was an embattled identity. And in exactly the same way, from the time I first took to the wheelchair up to the present, the fact that I am physically handicapped has been in the background of my conscious thoughts. Busy though I might be with other matters and problems, it lingers as a shadow in the corner of my mind, waiting, ready to come out at any moment to fill my meditations. It is a Presence. I, too, had acquired an embattled identity, a sense of who and what I was that was no longer dominated by my past social attributes, but rather by my physical defects.

One of the more interesting parallels between the stigma of handicap and other forms of embattlement is a sensitivity to nomenclature. One must refer to Negroes as *blacks* today, a term that would have been insulting forty years ago, when the polite word was *colored*. Likewise, the term *lady* is now considered patronizing, and *girl* seems reserved for the prepubescent. It is not surprising, then, that many people in wheelchairs take offense at the brutally direct word *paralysis*, and I have heard spirited arguments over the relative meanings and virtues of *handicapped* and *disabled*. Words such as *crip* and *gimp* are forbidden to the able-bodied, although they are used by the disabled among themselves; ethnic pejorative words are bandied about in the same way. I have treated *handicapped* and *disabled* as synonyms, for what I find most interesting about the debate over the words is the debate itself. It reveals a stance of defensiveness against belittlement that is seldom relaxed; it bespeaks a constant awareness of one's deficiencies. And in the process, even the vocabulary of disability has become emotionally charged. People have a hard time deciding what to say to the disabled, and their troubles are compounded by the fact that they are uncertain about what words to use.

In all the years since the onset of my illness, I have never consciously asked, "Why me?" I feel that this is a foolish question that assumes some cosmic sense of purpose and direction in the universe that simply does not exist. My outlook is quite fatalistic, an attitude that actually predisposes me to get all the pleasure out of life that I can, while I can. Nonetheless, though I may not brood over my impairment, it is always on my mind in spoken or unspoken form, and I believe this is true of all disabled people. It is a precondition of my plans and projects, a first premise of all my thoughts. Just as my former sense of embodiment remained taken for granted, positive, and unconscious, my sense of disembodiment is problematic, negative, and conscious. My identity has lost its stable moorings and has become contingent on a physical flaw.

This consuming consciousness of handicap even invades one's dreams. When I first became disabled, I was still walking, after a fashion, and I remained perfectly normal in my dreams. But as the years passed and I lost the ability to stand or walk, a curious change occurred. In every dream I start out walking and moving freely, often in perilous places; significantly, I am never in a wheelchair. I am climbing high on the mast of a ship in rough seas—something I did occasionally in an earlier incarnation— or I am on a ladder, painting a house. But in the middle of the dream, I remember that I can't walk, at which point I falter and fall. The dream is a perfect enactment of failure of power, the realization that what most men unconsciously fear had in fact happened to me. In other dreams, I am just walking about aimlessly when suddenly I remember my disability. Sometimes I sit down, but often I just stand puzzled until I awaken, the dream dissolves, the room comes into view, and I return to the reality that my paralysis is not a transient thing—it is an awakening much like that of Gregor Samsa in Kafka's *The Metamorphosis*. But perhaps more significant than the content of my dreams is the fact that since 1978 I have never once dreamed of anything else. Even in sleep, disability keeps its tyrannical hold over the mind.

The totality of the impact of serious physical impairment on conscious thought, as well as its firm implantation in the unconscious mind, gives disability a far stronger purchase on one's sense of who and what he is than do any social roles—even key ones such as age, occupation, and ethnicity. These can be manipulated, neutralized, and suspended, and in this way can become adjusted somewhat to each other. Moreover, each role can be played before a separate audience, allowing us to lead multiple lives. One cannot, however, shelve a disability or hide it from the world. A serious disability inundates all other claims to social standing, relegating to secondary status all the attainments of life, all other social roles, even sexuality. It is not a role; it is an identity, a dominant characteristic to which all social roles must be adjusted. And just as the paralytic cannot clear his mind of his impairment, society will not let him forget it.

Given the magnitude of this assault on the self, it is understandable that another major component of the subjective life of the handicapped is anger,[9] a disposition so diffuse and subtle, so carefully managed, that I became aware of it in myself only through writing this book. The anger of the disabled takes two forms. The first is an existential anger, a pervasive bitterness at one's fate, a hoarse and futile cry of rage against fortune. It is a sentiment fueled by the self-hate generated by unconscious shame and guilt, and it bears more than casual resemblance to the anger of America's black people. And, just as among blacks, it becomes expressed in hostility toward the dominant society, then toward people of one's own kind, and finally it is turned inward into an attack on the self. It is a very destructive emotion. In my own case, I have escaped its worst ravages only because my impairment has been so slow that I have been able to adjust to it mentally, and I am old enough to know that I am just a statistic, not the victim of a divine conspiracy. I suspect, although I lack conclusive data on this, that anger is much greater among those suddenly disabled and the young, for their impairment happens too quickly to permit assimilation, and it clouds an entire lifetime.

The other kind of anger is a situational one, a reaction to frustration or to perceived poor treatment. I have a good supply of this type. A paralytic may struggle to walk and become enraged when he cannot move his leg. Or a quadriplegic may pick up a cup of coffee with stiffened hands and drop it on his lap, precipitating an angry outburst. I had to give up spaghetti because I could no longer twirl it on my fork, and dinner would end for me in a sloppy mess. This would so upset me that I would lose my appetite. Or I may try unsuccessfully for a minute or so to pick up a paper from my desk or turn a page, casual maneuvers for most but a major challenge to me, because my fingers have lost both strength and dexterity. Such frustrations happen to me, and to other paralytics, several times a day. They are minor but cumulative, and they acquire special intensity from the more generalized existential anger often lurking below the surface.

The kind and virulence of the anger of the disabled vary greatly, for each person has a different history, but I have the impression that the depth and type of disability are critical. The extent of disablement obviously influences both existential and situational rage, but anger also seems to be most intense among people with communication disorders—primarily deaf-mutes and people with cerebral palsy and certain kinds of stroke. Most of us have watched the transparent suffering of the speech-impaired as they struggle to convey meaning to their agonized listeners. It is small wonder that the deaf form tightly circumscribed little communities, or that they occasionally explode into overt hostility at those who can hear and speak.

The anger of the disabled arises in the first place from their own lack of physical functions, but, as we will see, it is aggravated by their interaction with the able-bodied world. They daily suffer snub, avoidance, patronization, and occasional outright cruelty, and even when none of these occur, they sometimes imagine the affronts. But whatever the source of the grievance, the disabled have limited ways of showing it. Quadriplegics cannot stalk off in high (or low) dudgeon, nor can they even use body language. To make matters worse, as the price for normal relations, they must comfort others about their condition. They cannot show fear, sorrow, depression, sexuality, or anger, for this disturbs the able-bodied. The unsound of limb are permitted only to laugh. The rest of the emotions, including anger and the expression of hostility, must be bottled up, repressed, and allowed to simmer or be released in the backstage area of the home. This is where I let loose most of the day's frustrations and irritations, much to Yolanda's chagrin. But I never vent to her the full despair and foreboding I sometimes feel, and rarely even express it to myself. As for the rest of the world, I must sustain their faith in their own immunity by looking resolutely cheery. Have a nice day!

In summary, from my own experience and research and the work of others I have found that the four most far-reaching changes in the consciousness of the disabled are: lowered self-esteem; the invasion and occupation of thought by physical deficits; a strong undercurrent of anger; and the acquisition of a new, total, and undesirable identity. I can only liken the situation to a curious kind of "invasion of the body snatchers," in which the alien intruder and the old occupant coexist in mutual hostility in the same body. It is also a metamorphosis in the exact sense. One morning in the hospital, a nurse was washing me when she was called away by another nurse, who needed help in moving a patient. "I'll be right back," she said as she left, which all hospital denizens know is but a fond hope. She left me lying on my back without the call bell or the TV remote, the door was closed, and she was gone for a half hour. Wondering whether she had forgotten me, I tried to roll onto my side to reach the bell. But I was already quadriplegic, and, try as I might, I couldn't make it. I finally gave up and was almost immediately overcome by a claustrophobic panic, feeling trapped and immobile in my own body. I thought then of Kafka's giant bug, as it rocked from side to side, wiggling its useless legs, trying to get off its back—and I understood the story for the first time.

At the beginning of this chapter, I spoke of the feeling of aloneness, the desire to shrink from society into the inner recesses of the self, that invades the thoughts of the disabled—a feeling that I attributed in part to the deep physical tiredness that accompanies most debility and the formidable physical obstacles posed by the outside world. But we have added other elements to this urge to withdraw. The individual has also been alienated from his old, carefully nurtured, and closely guarded sense of self by a new, foreign, and unwelcome identity. And he becomes alienated from others by a double-barreled mechanism: Due to his depreciated self-image, he has a tendency to withdraw from his old associations into social isolation. And, as if in covert cooperation with this retreat, society—or at least American society—helps to wall him off.

The physical and emotional sequestering of the disabled is often dramatic. One quadriplegic man, married and the father of two children, told us that he never leaves the house and nobody visits their home, not even the friends of his children. He confessed to feelings of shame about his condition. I was struck by the similarity between that family and the one begotten by my father. Another quadriplegic we met attends college through a program that allows home study. Even though he is capable of leaving his house with help, he never does so, and instructors from the college have to meet him at his home. He is trying to break out of his shell, but he is not quite ready. Many disabled people blame their isolation on a hostile society, and often they are right. But there is also that powerful pull backward into the self. It is an urge that I have felt all my life, a centripetal force that is a universal feature of the emotional makeup of our species. Our lives are built upon a constant struggle between the need to reach out to others and a contrary urge to fall back into ourselves. Among the disabled, the inward pull becomes compelling, often irresistible, outlining in stark relief a human propensity that is often perceived only dimly.

The generality of my inquiry was brought home to me vividly one day while listening to a paper delivered by my colleague Katherine Newman. Newman has been doing important research on four groups of people who have experienced severe economic loss: divorced women, air-traffic controllers fired after going out on strike in 1981, laid-off blue-collar workers, and long-time-unemployed middle-management people. Newman described a pattern of consistent responses from all four groups. All experienced a deep sense of loss and went through a period of "mourning" quite similar to that reported among the traumatically disabled. Their feelings of depression were aggravated by a process of self-abasement, accentuated in the case of the divorcées by a sense of sexual inadequacy. Common to most members of the four groups was the idea that they somehow were responsible for the loss, that they had failed as providers. They felt culpable, even in cases where they clearly were innocent victims of impersonal economic circumstance. Here, too, the American ideology of success, combined with vestiges of Calvinism, takes the anger that should be aimed at the system and turns it inward upon the self. With their guilt came shame, and Newman's informants frequently surrendered to that sentiment by sharp curtailment of their social contacts. These tendencies often were reinforced by society's penchant for blaming the victim; their self-condemnation was joined by the censure of others. The stricken individual, and his or her family, withdraws into humiliation, and all tend to be avoided just as if a pox had visited them.

The psychological devastation wrought by unemployment has been studied for more than half a century, but Newman's brilliant exposition makes vividly clear the striking parallels between economic and physical disability, the despoilment of identity that is the common fallout of the damaged self. It is a commentary on the importance of economic status in America that downward mobility fosters the same social and psychological results as crippling. And it is worth noting at this point that the social and emotional ravages of physical disability often are magnified by the individual's loss of livelihood. My own case is a rare exception.

Most of Newman's subjects will eventually make their way back to some kind of economic viability, and here they part company with the handicapped, to whom something more devastating has happened. The disabled have become changed in the minds of the rest of society into a kind of quasi-human. In only a few months, I had moved subtly from the center of my society to its perimeter. I had acquired a new identity that was contingent on my defects and that either compromised or radically altered my prior claims to personhood. In my middle age, I had become a changeling, the lot of all disabled people. They are afflicted with a malady of the body that is translated into a cancer within the self and a disease of social relationships. They have experienced a transformation of the essential condition of their being in the world. They have become aliens, even exiles, in their own lands.

5

ENCOUNTERS

Our own body is in the world as the heart is in the organism; it keeps the visible spectacle constantly alive, it breathes life into it and sustains it inwardly, and with it forms a system.
—Maurice Merleau-Ponty,
*The Phenomenology of
Perception*

The recently disabled paralytic faces the world with a changed body and an altered identity—which even by itself would make his reentry into society a delicate and chancy matter. But his future is made even more perilous by the treatment given him by the nondisabled, including some of his oldest friends and associates, and even family members. Although this varies considerably from one situation to another, there is a clear pattern in the United States, and in many other countries, of prejudice toward the disabled and debasement of their social status, which find their most extreme expressions in avoidance, fear, and outright hostility. As Erving Goffman noted in his landmark 1963 book, *Stigma: Notes on the Management of Spoiled Identity*, the disabled occupy the same devalued status as ex-convicts, certain ethnic and racial minorities, and the mentally ill, among others.[1]

Whatever the physically impaired person may think of himself, he is attributed a negative identity by society, and much of his social life is a struggle against this imposed image. It is for this reason that we can say that stigmatization is less a by-product of disability than its substance. The greatest impediment to a person's taking full part in his society are not his physical flaws, but rather the tissue of myths, fears, and misunderstandings that society attaches to them.

To understand why this is so, it is necessary to consider some of the central themes of American culture, especially our attitudes toward the body. The body is so important in American symbolism that most of us, including anthropologists, do not even realize that its care and nurture have changed from practicality to fetishism. This is not simply a phenomenon of the past decade. In 1956, the anthropologist Horace Miner wrote a wonderful, tongue-in-cheek essay on a people he called the Nacirema (*American* spelled backward).[2] With unerring accuracy, Miner described the bathroom as a religious center where the inhabitants make their ritual ablutions in a cult of the body beautiful. Since that time, the bathroom, surely one of the capstones of our culture, has evolved even further; those of the rich now have huge, sunken tubs with built-in Jacuzzis, and the profane toilet bowl is segregated from the sacred bathing area. Whatever else may be said about Americans, they are a fairly clean people.

But the body must be more than clean; it must have a certain shape. The anthropologist Marvin Harris notes that whereas corpulence used to be an indication of wealth and prestige— Diamond Jim Brady and J. P. Morgan come to mind—it now is a sign of lower-class status and an overly rich diet. The reigning beauties of the Gay Nineties would be buxom, even fat, by today's standards, for our twentieth-century ideals of beauty have evolved from Lillian Russell to Marilyn Monroe to Twiggy. Today's bodies must be lean and muscular, an injunction that is almost as binding on females as it is on males. The feminine ideal has shifted from soft curves to hard bodies.

113

And how does one attain the body beautiful of the 1980s? By exercise, diet, and other mortifications of the flesh. If you want the right kind of body—presumably the passport to romantic love and economic success—then you must get out there and jog several miles a day. The craze for jogging soon turned to running frenzy, for it has been deemed necessary to get the heart pumping hard. Another route to the ideal body is the health club, and millions of Americans have become members of thousands of spas with such features as swimming, squash, racquetball, weight lifting, sauna, whirlpool, massage, and aerobics. This injunction to exercise is especially intense among the upwardly mobile middle class, although it also has become part of the lifestyle of the working class. None of this is to deny that exercise is good and healthful. What interests the anthropologist, however, is that its practice in contemporary America extends beyond rational self-interest to zealotry. And the pursuit of the slim, well-muscled body is not only an aesthetic matter, but also a moral imperative.

The morality of the good body is manifest in the message pounded out daily in television commercials that "self-improvement" means attaining physical fitness, an even more mindless activity than the transcendental meditation of the 1970s. Obesity is regarded as punishment for sloth and weak will, and this is nowhere more evident than in the American preoccupation with diet. "You are what you eat" and "the body is the temple of the mind" are among the platitudes by which we live, although to the extent that one can fathom their meanings, both are nonsense. Nonetheless, if the body is a sacred zone, one must be careful about what one puts into it. Health-food stores now abound, organically grown products are popular, and vegetarianism is on the rise. Fasting and self-inflicted physical punishment are the modern-day equivalents of medieval flagellantism. They are religious rituals, part of the immortality project of a secularized middle class that no longer believes in redemption of the soul and has turned instead to redemption of the body.

Both men and women have found that a youthful appearance is a considerable asset in the business world—quite the opposite of the situation in Japan, where maturity commands respect. In the United States, cosmetic surgery for women has been on the increase in recent years, but even more dramatic has been its exponential growth among men. The emulation of youth has extended itself to fashion as well. During the 1960s, young people began to let their hair grow long as an act of separation and defiance—only to find that in a few short years, long hair had spread to stock brokers and advertising men, who now have their hair cut (or, rather, "styled") at exclusive salons. Clothing has become more youthfully casual, too. Teachers wear the same blue jeans as their students and suits have a less severe cut. To accentuate youthful beauty, men now use cosmetics, something that would have cost them their masculine credentials in my benighted era. But *these* are he-man products. To cite just one example, designer Ralph Lauren has "designed" a men's cologne called "Chaps" (as in the chaps worn by cowboys), a concoction, the TV commercial assures us, that expresses the masculine values of the Old West. It's grand hokum like this that makes the present day and age so delicious for us anthropologists.

Much of the ideal America purveyed by the mass media applies only to the upwardly mobile, but on close examination it distorts even their circumstances. In reality, this is a country in which the gulf between the haves and the have-nots is large and growing, and in which the general standard of living has been inching downward ever since the mid-1960s. This stark truth is based on ample evidence that the purchasing power of the average family has declined almost five percent since 1971; the drop would have been far steeper but for the dramatic increase in the number of working wives. The country, however, operates under the illusion that we are living better. After all, secretaries and factory workers travel to Europe on vacation now, something that once only the wealthy could do. But this is merely because jet aircraft—flying cattle cars carrying nearly 400 heads per trip—now make the fare to London as cheap in constant dollars as a trip to Niagara Falls was in 1950. But most of today's younger tourists in Europe cannot afford to buy a house or have more than one or two children, something their parents were able to do in 1950. And they did so on one income. Are we really living better?

America is a land of shrinking resources and families, a society whose culture glorifies the body beautiful and youthfulness, while barely tolerating youths. It is small wonder that it harbors people increasingly turned in on themselves in rampant narcissism. To make matters worse, the increased affluence of the upper middle class and upper class is offset by the growing despair of the lower class. The economic plight of their men is eroding the black family, and once proudly independent blue-collar workers now line up in soup kitchens, their jobs sent overseas by American capital. Our cities are littered with homeless people sleeping in bus stations and doorways, rummaging for food in garbage cans, abandoned by a society that dodges responsibility by telling itself that such people choose to live that way. The successful simply shrug their shoulders and say, "I'm all right, Jack." Our cities present scenes that are almost reminiscent of Calcutta. Europeans are appalled, but to most urban Americans the homeless have become invisible. They walk around the human rag piles, they avert their eyes, and they maintain the myth that they dwell in what some politicians have called "a shining city on a hill."

They do the same thing with the disabled.

The kind of culture the handicapped American must face is just as much a part of the environs of his disability as his wheelchair. It hardly needs saying that the disabled, individually and as a group, contravene all the values of youth, virility, activity, and physical beauty that Americans cherish, however little most individuals may realize them. Most handicapped people, myself included, sense that others resent them for this reason: We are subverters of an American Ideal, just as the poor are betrayers of

116

the American Dream. And to the extent that we depart from the ideal, we become ugly and repulsive to the able-bodied. People recoil from us, especially when there is facial damage or bodily distortion. The disabled serve as constant, visible reminders to the able-bodied that the society they live in is shot through with inequity and suffering, that they live in a counterfeit paradise, that they too are vulnerable. We represent a fearsome possibility.

What makes the disabled particularly threatening are the psychological mechanisms of projection and identification by which people impute their feelings, plans, and motives to others and incorporate those of the others as their own. Through these processes, the disabled arouse in the able-bodied fear that impairment could happen to them and, among relatives and friends, guilt that it hasn't. In their excellent 1979 book, *The Unexpected Minority*, John Gliedman and William Roth write that the disabled person becomes the Other—a living symbol of failure, frailty, and emasculation; a counterpoint to normality; a figure whose very humanity is questionable.[3] So prevalent is the dehumanization of the disabled that I have heard it inveighed against repeatedly by disabled people; and the fact that it is not just an American trait is bespoken by the title of Norwegian author Finn Carling's book *And Yet We Are Human*.[4]

It is clear that the disabled arouse primordial sentiments in people throughout the world, but lack of comparative data makes it difficult to say where, how, and why. We do know that the stigma of disability is much worse in Japan than in the United States, and that the aura of contamination that often surrounds the disabled becomes attached to other members of the family. One Japanese author has written of the mother of a congenitally deformed child, who felt such shame and despair that she attempted suicide.[5] The author attributes this severity to a Japanese belief that the individual somehow is to blame for his or her own misfortunes and to the cultural and ethnic homogeneity of the population. In contrast, northern European attitudes toward the disabled are more relaxed and positive than

Japan's or our own, and their rehabilitation programs are correspondingly more advanced. Another accommodation to disability was related to me by my Columbia colleague Morton Klass, who watched one day as a group of men in India teased a blind man. Klass at first thought this to be cruel, but then he realized that the banter was a way of including the blind man in the group; it was a joking relationship of the same type that anthropologists have found among in-laws in some primitive societies. Significantly, this usually takes the form of avoidance of parents-in-law and joking toward brothers- and sisters-in-law, people one may marry; the banter thus can be seen as an alternative way of putting distance between people. In this light, Indian joking toward the disabled is a simple transposition of American avoidance of them. I suspect that future studies will reveal that in lands where poverty and disease are rampant, the disabled will not be as excluded from social life as they are in the United States, but it probably will also come out that they always receive special treatment.

Returning to American culture, there is deep and uneasy ambivalence in relations between the able-bodied and the disabled, for how is one supposed to act toward a quasi-human, a person who literally arouses fear and loathing? These are sentiments that must remain hidden at all costs, for they fly in the teeth of values that dictate concern for and kindness to the handicapped. Social encounters are always tricky games, sparring matches in which each party tries to guess what the reaction of the other will be, and the game is made more difficult when one person has no inkling of what the other is like, such as in the case of a foreigner—or a paralyzed person. To the extent that people look on the disabled as an alien species, they cannot anticipate their reactions; the disabled individual falls outside the ken of normal expectations, and the able-bodied are left not knowing what to say to him or her. One way out of the dilemma is to refrain from establishing any contact at all. This can be done with people in wheelchairs simply by physical avoidance, an easy solution for a person with two working legs.

So pronounced and widespread is the aversion of eyes and setting of physical distance that I have never met a disabled person who has not commented on it, and the literature on disability is too replete on this point to be worth citing. The disabled often say, "People act like it was catching." This exact expression was uttered by a Japanese woman maimed by the atomic bomb at Hiroshima, in describing the fact that nobody ever visits the victims. It is worth noting that people with cancer and those suffering loss through death or divorce are similarly given wide berth; they have all suffered a contamination of identity. Malignancy and mourning are, however, transient conditions. Paralytic disability is not.

Erving Goffman, who was a sociologist by training but an anthropologist in practice, wrote in 1956 that the very core and starting premise of all social interaction is the establishment by the people involved of stances of deference and demeanor.[6] Each party must comport himself or herself as a person of worth and substance, and each must put social space and distance around the self. The other, in turn, respects this demeanor by according deference. The extent of this mutual respect varies, of course, with the situation and the people involved, and the way in which it is expressed is an artifact of culture. It occurs through a subconscious grammar of gesture and verbal nuance, a language so subtle that it escapes the awareness of both user and hearer, except when it is withheld—as it so often is for the physically impaired.

Whatever the cultural and situational variations, a broad segment of the general public handles relations with the disabled by a partial withdrawal of deference. They do this unconsciously and in a variety of ways. The violinist Itzhak Perlman, who suffers from the aftereffects of polio, says that when he is pushed up to an airline counter in a wheelchair, the clerk commonly asks his attendant, "Where is he going?" This has happened to me on numerous occasions, and I was reminded again that it is not just an American phenomenon when a waiter in a Korean restaurant handed out three menus to our party of four; I called him back and told him I too knew how to read. It did little good, for he did the same thing two months later. People also speak loudly to the blind on the assumption that they are deaf as well. Most disabled people can swap endless anecdotes on this theme.

This kind of denigration is a universal complaint of the handicapped: "You'd think I was retarded, too." When the able-bodied are forced into confrontation with the disabled—that is, when they cannot escape—they often cope with the threat by treating the disabled as minors or as incompetent, withholding deference and thereby depriving them of their due as fellow humans. They are also differentiating themselves from the disabled person by asserting their superiority, as if this would somehow make them less vulnerable to a similar fate. To make matters worse, the disabled, particularly the deformed, are sometimes seen as evil, as in Shakespeare's *Richard III* or Victor Hugo's *Hunchback of Notre Dame*. This may well be a projection of the inner hostility of the able-bodied toward the handicapped, a sentiment that surely exists, however seldom it is shown. And so it is that physical impairment is generalized even to character, a process called "spread" by social psychologist Beatrice Wright.[7]

Most Americans, including medical people, carry around inside their heads a set of notions about the social position of the handicapped. Whatever its other qualities, this mindset does not place the disabled in the social mainstream, but rather on the periphery, pensioned off and largely out of sight. Even the hospital personnel found me anomalous, for not only was I fully employed, but I was doing research in their own area of expertise. One social worker asked me, "What *was* your occupation?" It is not that they begrudged me my ability to keep working. Quite to the contrary, they were honestly pleased by it. It simply made me a very different social type, a special case. My own doctors know this, of course, and they have often corrected a medical colleague's recommendations by saying, "He can't do that, he's a working man." But he's a quadriplegic working man, and this one trait now defines me. It need not always be so, however. Consider Franklin D. Roosevelt's triumph over such total categorization. He muted his disability by always standing to deliver speeches and never allowing himself to be photographed in his wheelchair.

The disabled person must make an extra effort to establish his status as an autonomous, worthy individual, but the reaction of the other party may totally undercut these pretensions through some thoughtless act or omission. Even if the able-bodied person is making a conscious attempt to pay deference to the disabled party, he must struggle against the underlying ambiguity of the encounter, the lack of clear cultural guidelines on how to behave, and perhaps his own sense of revulsion. This often lends an atmosphere of forced artificiality to an occasion. It can either go flat through resort to formality, or into hyperbole, as in fake joviality, pretenses of humor, or effusive friendliness. Meetings between the able and the disabled can indeed be awkward, tense, and indeterminate affairs.

The distortion of social scenes by some salient and anomalous trait of one of the parties is not limited to the handicapped. A friend once told me of a party he attended at which a female guest wore a wide-mesh net dress and nothing else. She was, for all practical purposes, naked. Now there are some social circles in which the young woman would have been at serious risk—if not from the men, then from the other women. This, however, was a gathering of middle-class intellectuals, people who pride themselves on their urbanity and sophistication, their lack of sexism. Nobody spoke of the dress or its wearer, and the men took care not to look at her. When a man conversed with her, he would train his eyes on her face in a deliberate effort to keep his glance from dropping to where it wanted to go. Not surprisingly, nobody spoke to her much. My friend commented that the party was a flop. Its tone remained formal, its conviviality was forced, and it broke up early.

The woman's see-through dress became the unspoken center of social dialogue, it subverted all other party activity, and it distorted all other relationships. Small wonder that people fled.

121

In what I believe is the best essay on the sociology of disability yet published, Fred Davis wrote that the same thing happens during encounters, especially initial ones, between the able-bodied and the handicapped.[8] My own research and personal experience confirm his thesis. Just as one's identity as a disabled person is paramount in his own mind, and the impairment an axiom for his actions, so too is the other's discernment of the handicapped individual overwhelmed by the flaw. This one obvious fact, the disabled person's radical bodily difference, his departure from the human standard, dominates the thoughts of the other and may even repel him. But these are thoughts that can barely be articulated, let alone voiced.

The disability—paraplegia, blindness, or whatever it may be—is at center stage, in the forefront of the consciousness of both parties, and both must take steps to normalize the meeting, a process that Davis calls "deviance disavowal." The participants try to conduct themselves as if nothing were amiss, as if there were no hidden agenda. Several different scenarios are possible aside from avoidance and patronization. One technique is to make a brief allusion to it at the outset, as if to say, "There, that's on the table and out in the open; now let's get on with our business." This line of action is usually set in motion by the impaired person, who has to become an expert at putting others at their ease. He does this, as I have said, by cheerful demeanor; anything else would make the other person run away. But every now and then the disabled party lets the other one writhe. It's a good way of getting rid of unwanted company.

Davis correctly states that in these primal scenes the hidden agenda, the dominant yet unspoken flaw, distorts sociability. The able-bodied one is worried that he might say something hurtful, and he tiptoes into the encounter as if he were walking through a mine field. The disabled one knows what the other is thinking about, and the latter knows that this awareness is known; each knows that the other knows that he knows that he knows . . . as in a hall of mirrors. But these, again, are Coney Island mirrors, which both reflect and distort, and the normalization process operates on a bed of quicksand, always in

122

danger of being engulfed. Georg Simmel, one of the founders of modern sociology, once wrote that all social encounters threaten imminent disaster, and those with the handicapped are extreme illustrations of this general truth. Simmel also wrote that social action is predicated upon a "teleologically determined non-knowledge of one another," [9] meaning that if people had clear, accurate personal information about each other and observed total honesty, it would destroy sociability and make human society impossible. We can add that not only does each party withhold information from the other, but also each distorts and embellishes those nuggets of fact that are released. And each suspends disbelief and half-accepts these little white lies as a necessary price for getting along. What makes the interaction of the disabled with the able-bodied so fascinating is that it rests not only on little fibs but also on a big lie—that the physical deficiency makes no difference. It does, and in the uneasy interaction between the two people, misunderstandings become magnified and sociability inverted.

The underground salience of disability was neatly illustrated by a friend of mine who uses crutches because of childhood polio. He boarded an airplane, settled in a seat, and gave his crutches to the flight attendant. A woman sat down beside him, and the two started a friendly conversation that lasted until they landed. When the plane reached the terminal, the attendant returned with his crutches. Seeing them, the woman became flustered and embarrassed, muttered a quick good-bye, and debarked hastily. She had not learned of the great mortgage on his identity until too late. I am sure that she wondered for the rest of the day whether she had said something "wrong," and it is clear that she would have behaved differently had she known. Her solution was to flee.

Davis was concerned primarily with initial encounters, occasions of which one able-bodied informant said, "The first problem is where to direct your eyes." Davis's observations also apply to meetings with old associates, however. These can be more trying than encounters between strangers, for a conscious effort must be made to maintain previously established social roles, an attempt that runs afoul of the disabled person's new identity and the other's aversion, guilt, and fear. The former is aware of the inner turmoil of his friend, and the latter feels a sense of estrangement, as if his old associate, now stricken, had gone somewhere else—which he has. Their relationship must be redefined, which is often a harder job than forging a new one. As a result, the recently disabled often drop, or are dropped by, their old friends, kinfolk, and even their spouses.

The social circles of the disabled are foreshortened and shrunken, their associates diminished in number and often drawn from different social strata. Because of my truncated social world, I no longer go to anthropological conventions (which I have regarded as a waste of time ever since I quit drinking), with a consequent attrition of my professional circle, just as my withdrawal from broader university life has embedded me in my department and intensified my relations with colleagues and students. The same thing has happened in my suburban community, where I had made a hobby of local politics and had a large circle of friends. At first they rallied around me, but as my illness progressed they began to drop out of sight, which was partially my fault, for I seldom invited them to visit. Isolation is a two-way street.

The fact that my isolation was not solely my fault, however, was brought home forcibly when a friend with muscular dystrophy and I organized a well-publicized program on disability at the town library. Naively, I had expected a large crowd of friends, neighbors, and political cronies, but only eight people showed up. I had failed completely to realize how much people were repelled by the subject, and how ambivalent their feelings had become toward me. The incident made me resentful, however, and I withdrew into a small circle of close friends who have become at ease with my disability. I realize now that in doing so I cut myself off from a number of other good people who are honestly puzzled about how to approach me. Some admit that they cannot stand what has happened to me, so they stay away. One not-so-close friend reportedly said, "What a shame. He *was* such a nice guy." The speaker's use of the past tense was not just an accident, for rumors surface every now and then that I am at death's door. I savor these incidents as raw data, for they harbor a metaphoric truth: What died was the old social me.

The avoidance of the disabled cannot be said to be merely a result of ignorance and fear of the unfamiliar. The equanimity of the medical profession was sorely strained in 1982 by a paper in the prestigious *New England Journal of Medicine* by Dr. David Rabin, at that time a faculty member at the Vanderbilt University Medical School.[10] Rabin, now deceased, had amyotrophic lateral sclerosis, which is always fatal, and this became widely known. He expected sympathetic understanding from his colleagues, but instead he found that they were avoiding him. This became more marked as the disease progressed, until one day he slipped and fell in the hospital, at which point a nearby doctor looked away. In general, he received more support and assistance from subordinates than from his peers. A great barrier separates physicians from their terminal patients, and Rabin had breached this wall. The profession consolidated ranks, and, by isolating Rabin, sealed off the mixing of healer and doomed. My anthropological colleagues have behaved much better than that, but then human variety is the stuff of their craft.

The reduction of one's social universe is a qualitative matter as well as a quantitative one. The handicapped often go on to make new contacts, and researchers uniformly report that these tend to be with people of lower social standing than the old peer group.[11] They befriend other disabled people whom they meet through clubs and church organizations, they associate with other unemployed people, and they feel ill at ease with the affluent—who are also among the first to avoid them. One of Fred Davis's able-bodied, upper-middle-class informants said that it was hard to assimilate serious impairment happening to "people of our status."[12] This echoes a small episode in my own experience. I was attending a meeting of a disability rights organization and spotted across the room a state legislator whom I knew from my pre-wheelchair politicking. He at first didn't recognize me, for the chair served as a kind of disguise, but after he had placed me personally, and socially, he said that he hadn't expected to find "a person of your quality" at such a gathering. Such things are supposed to happen only to life's losers.

Owing to my age and long-established position, I was able to maintain my closest and most valued ties, although I attenuated some peripheral ones. I did not shift my associations down the social scale, although my research between 1980 and 1983 on social relations of the disabled did bring me into contact with many economically marginal people. I must admit, however, to feeling more uneasy in the first stages of this research than among Amazonian Indians—not *despite* the fact that I too am disabled, but *because* of it. I coped with the rather unexpected discomfort I felt when among others in wheelchairs by assuming the investigator's fiction that he is an independent and objective observer, *in* but not *of* the group he is studying. This neat separation of subject and object is never valid, and it was doubly false in my case. I had cherished a personal myth of almost-normalcy and took pride in my productive life; I was not yet ready to identify fully with the disabled. The same reaction was reported by Finn Carling, who wrote, "As far back as I can remember, I withdrew very sharply from any contact with other cripples." [13] Research among the motor-handicapped and participation in their organizations forced me to see myself in their lives, and this left me feeling that my own status was insecure and threatened. The research was too close to home for comfort—and I had learned a valuable lesson about the relationship of social standing to disability. I had also learned a great deal about myself. All anthropological research involves a process of self-discovery, and my experience among the disabled was often painful.

Not long after I took up life in the wheelchair, I began to notice other curious shifts and nuances in my social world. After a dentist patted me on the head in 1980, I never returned to his office. But undergraduate students often would touch my arm or shoulder lightly when taking leave of me, something they never did in my walking days, and I found this pleasant. Why? The dentist was putting me in my place and treating me as one would a child, but the students were affirming a bond. They were reaching over a wall and asserting that they were on my side. I was a middle-aged professor and just as great an exam threat to them as any other instructor, but my physical impairment brought them closer to me because I was less imposing to them socially. As for the graduate students, it was not until after the onset of my disability that many began calling me by my first name, which was also a demonstration of closeness, not over-familiarity.

The same thing happened in my contacts with black people. I used to be invisible to black campus policemen, who often greeted a black colleague with whom I was walking by saying deliberately and clearly in the singular, "Hello, Professor, how are you today?" They now know who I am and say hello. I am now a white man who is worse off than they are, and my subtle loss of public standing brings me closer to their own status. We share a common position on the periphery of society—we are fellow Outsiders.

During my first couple of years in the wheelchair, I noticed that men and women responded to me differently. My peer group of middle-aged, middle-class males seemed most menaced by my disability, probably because they identify most closely with me. On the other hand, I found that my relations with most women of all ages have become more relaxed and open; they are at once more solicitous than men and more at ease in my company. I noticed, too, that when I got on the elevator with a woman, she often would greet me or start a conversation; in my walking days, we both would have stared silently at the floor indicator. The same thing would happen when I was being wheeled across campus. As a little experiment, I would look at the face of an approaching woman until I caught her eye. At this point, the woman ordinarily should look away, but in most cases she would instead lock glances and nod or smile. The exchange of eye contact was an opening of the self, an acknowledgment of the other, a meeting without closure.

I found this new openness refreshing and agreeable, for despite being totally monogamous, I have always enjoyed the company of women. They are generally nicer people than men, although I should note quickly that there are many exceptions to this blatantly glib and unanthropological generalization. But what did this reaction tell me of relations between the sexes? It confirmed Freud's thesis that men and women are separated by a wall of antagonism, and that both sexes have elaborate, largely unconscious, mechanisms of defense against the other. And this was not merely a feature of Freud's Vienna. Yolanda and I found vivid evidence of this antagonism among the Mundurucu, where it was embedded in a living arrangement in which men and women slept in separate houses.[14] I argued in another paper that women everywhere commonly defend themselves by the observance of decorum and restraint; they are trained from childhood to say no, and to see every male as a potential threat.[15] And despite all the recent changes in sexual standards, I believe that this is still the prevailing attitude in the United States.

The reason, then, for my new ease with women was that I was no longer a source of danger. After all, even if I wanted to pursue a woman, she could easily outrun me. Women were in total command of this aspect of our relations. One might protest that an aging, respectable professor is not much of a threat anyhow, except to himself, but this would overlook the fact that female defensiveness is based on a deeply ingrained, largely preconscious anxiety. It is less an active element in gender relations than an unspoken premise. Most of us do not realize the depth of this disposition, and I became aware of it only because my position as a man had been submerged by my new identity as a paraplegic.

Other researchers have also noted that women relate more easily to the disabled,[16] and some have attributed this to the traditional female role in nurturing and in the care of the ill. This may indeed be a factor, but I doubt whether it is the major one. Women went into the nursing field in the past because it was one of the best jobs open to them. Now they prefer to become physicians or bankers. As for their having a special affinity for nursing, I can only remark that some of the finest nursing care I have received has been from men—and some of the worst has been from women. Rather than impute some kind of mothering instinct to women, it is much more to the point to note again that disability is a great leveler. It forecloses an ancient power struggle and puts an end to "male superiority."

Erving Goffman's *Stigma* had great influence on the sociological study of disability by providing a common framework within which the handicapped, criminals, and certain minority groups could be seen as sharing a common lot: They are all outsiders, deviants from social norms. There are, however, problems with this framework. First of all, it throws into one pot people who are deliberate violators of legal or moral standards and persons who are in no way to blame for their stigmatized state. A person chooses to follow a life of crime, but nobody asks to be born a black, and certainly nobody wants to become a quadriplegic. These stations in life are visited upon people by inheritance or bad luck, not through choice. This, of course, does not prevent others from blaming the victim, and all too many benighted whites look on blacks as lazy and unintelligent people who prefer welfare and crime to working for a living. Even the disabled are often vaguely blamed for their condition, or at least for not achieving maximum recovery. And as sure proof that they bear stigmatized identities, physical impairment is looked upon as something that does not happen to respectable people. The blind are folks who make brooms in sheltered workshops, or who sit on street corners with tin cups. They certainly do not belong among the upwardly mobile.

The handicapped and the dark of skin differ from the felon in degree of culpability, but they differ from each other as well. Racial prejudice in the United States has deep historic and economic roots, for blacks and Hispanic migrants have served for centuries as a pool of cheap labor, a position they now share increasingly with women—who could well be included among the stigmatized. It pays to keep them down. There are, however, no strong economic reasons for the systematic exclusion and abasement of the physically handicapped, except for the minor fact that they often are supported and cared for at public expense. Otherwise, it is difficult to understand how discrimination against them can serve any significant social function. Nonetheless, research indicates that people who harbor hostility toward the disabled are statistically more likely to be prejudiced toward minorities.[17] There is an element of plain nastiness in all this; bigotry observes no boundaries.

In addition to the structural differences between race and handicap, a different scale of values and emotional responses applies to the disabled. People are socialized to racial prejudice—they are taught to hate Jews and blacks—but not to discriminate against the disabled. Despite this, the physically impaired often arouse, in varying degrees, revulsion, fear, and outright hostility—sentiments that appear to be spontaneous and "natural" because they seem to violate our values and upbringing. But *do* they? Children are quite understandably curious about disabled people and often stare at them, only to have their parents yank their arms and say, "Don't look." Nothing could better communicate to a child a sense of horror for disability; the condition is so terrible that one cannot speak about it or even look at it. Children are in many such ways taught to regard impairment with a loathing far beyond that of racial prejudice. It is a sentiment that reinforces the fear that this could happen to them.

As for the injunction that the handicapped should be helped, we do this from a safe distance, by contributing to such organizations as the March of Dimes and the Muscular Dystrophy Association or by dropping coins in a beggar's cup. In this way, the able-bodied lull their consciences without getting too close; they stress their own separation and intactness by an act of charity. These contradictory reactions of kindness and rejection help make the treatment of the disabled the arena of enormous conflicts of values.

Viewing disability as a subtype of deviancy confuses many

issues, leading to a theoretical dead end for social scientists. During the course of our research, my associates and I found it much more profitable to look upon disability in a different framework, one that simultaneously universalizes the condition and preserves its uniqueness. We treated disability as a form of *liminality*—a concept I introduced when discussing my state of mind after neurosurgery.[18] It is also closely related to the rites of passage, which I mentioned in chapter 3 in connection with the award dinner that marked my return to Columbia. Initiation rituals have the purpose of involving the community in the transformation of an individual from one position in society to another. They typically do this in three phases: isolation and instruction of the initiate, ritual emergence, and reincorporation into society in the new role. It is during the transitional phase from isolation to emergence that the person is said to be in a liminal state—literally, at the threshold—a kind of social limbo in which he is left standing outside the formal social system.

We owe a great deal of our understanding of ritual to Arnold van Gennep and Émile Durkheim, and to Durkheim's students Henri Hubert and Marcel Mauss, but it is the anthropologist Victor Turner who has done most to bring their ideas into line with modern cultural and social theory. The title of one of his essays, "Betwixt and Between," is actually a neat description of the ambiguous position of the disabled in American life.[19] The long-term physically impaired are neither sick nor well, neither dead nor fully alive, neither out of society nor wholly in it. They are human beings but their bodies are warped or malfunctioning, leaving their full humanity in doubt. They are not ill, for illness is transitional to either death or recovery. Indeed, illness is a fine example of a nonreligious, nonceremonial liminal condition. The sick person lives in a state of social suspension until he or she gets better. The disabled spend a lifetime in a similar suspended state. They are neither fish nor fowl; they exist in partial isolation from society as undefined, ambiguous people.

This undefined quality, an existential departure from normality, contributes to the widespread aversion to the disabled reported by researchers. The anthropologist Mary Douglas wrote in her 1966 book *Purity and Danger* that cultural symbolism sorts out conventional reality into tidy categories, and that departures from these neat classifications are regarded in many cultures as dangerous.[20] She argues that the anomaly posed by the pig, an animal with cloven hoofs that doesn't chew its cud, is the reason for the Hebrew taboo on the eating of its flesh. Lack of clarity means lack of cleanliness—pork, therefore, is polluting and must be avoided. The permanently disabled also fall into the category of the contaminated—and for much the same reason. They are anomalies, like deeply spastic people or the so-called Elephant Man, who had the dubious honor of being the most facially deformed person of his time. In the often-brutal argot of our own age, the severely disabled are "downers"; they depress people and are best kept away from places of relaxation and enjoyment. They also "gross out" ordinary folks, meaning that they cause disgust or revulsion. Not everybody reacts this way, of course, but it still is quite common. Until very recently, restaurants occasionally turned away the obviously handicapped, an attitude that changed only after concerted efforts were made to reeducate the public. Some disabilities disturb the able-bodied more than others. There is a hierarchy of devaluation that varies with severity and type of disability. At the bottom of the scale are persons with facial disfigurement or marked body distortion; wheelchairs are somewhere in the middle. The main criterion for placement seems to be degree of departure from the standard human form.

One can add to Douglas's theory Claude Lévi-Strauss's idea that the greatest of all binary distinctions in human thought is the separation of nature and culture. Placed in the framework of this grand dualism, the physical impairment is an infringement by nature, an intrusion that undercuts one's status as a bearer of culture. This same process is at work in societies that isolate women during their menstrual periods or after childbirth. This, finally, is what makes disability so different from other kinds of "deviance." It is not just a departure from the moral code, but a

distortion of conventional classification and knowing. The contamination of the handicapped by nature joins with the logical
anomaly posed by their bodies to compromise their very
humanness.

Turner writes that persons undergoing ritual changes of status
"are at once no longer classified and not yet classified."[21] They
have lost their old status and have not yet acquired a new one.
This leaves others uncertain about how to act toward them, and
here we have an echo of the recurrent quandary of what to do
about the disabled. This indeterminacy can be resolved by segre-
gation or avoidance of such liminal people—they become ritu-
ally polluted in Douglas's sense. In simple, primitive societies,
puberty-rite initiates may be sequestered for weeks, months, or
even years, a removal that in modern, complex societies is accomplished by such mild measures as after-school religious instruction and the honeymoon. A far more severe form of
isolation is effected by the confinement of the disabled to hospitals and nursing homes, or by their inability to leave their dwellings because of the physical barriers of curbs, steps, and
inaccessible public transit.

There are other striking parallels between the disabled and
initiates. Turner writes that "between instructors and neophytes
there is often complete authority and complete submission;
among neophytes there is often complete equality."[22] This certainly describes the authoritarian and tutelary role of medical
people, who serve the same purpose in rehabilitation wards as
do tribal elders in bush schools. The equality of the neophytes is
also present. Hospitals strip people of their previous identities
and reduce them to the amorphous status of "patient," and anyone who has spent long spells in these establishments knows
that the patients usually interact as equals, ignoring each other's
prior social distinctions.

This parity of rank also occurs outside the hospital among the
disabled. During the past ten years, I have participated in a number of organizations of the handicapped and attended countless
meetings, and I have been struck with the egalitarian atmo-

sphere that prevails. This equality has been extended to me despite the fact that I often have been the oldest person present and almost always the holder of the most prestigious professional position. Nobody, however, has called me "doctor" or "professor" or even "mister"; they have used only first names. There has been some deference to the fact that I was more informed about disability than most of them, but it has given me little authority. In fact, I often have been chagrined to find that many of my opinions have been shrugged off—a blow to the ego of somebody accustomed to having his audiences take notes on everything he says. What was even more threatening to my professorial stuffiness, however, was the fact that much more serious attention was paid to able-bodied, outside "experts." The disabled thus engage in the same invidious distinctions and practices they ostensibly deplore.

Our shared identities as disabled people override the old hierarchies of age, education, and occupation, and they wash out many sex-role barriers as well. I first noted this when undergoing physical therapy in 1976. Immediately after being introduced to a young woman with a partially paralyzed leg, I asked her, "How is your therapy going?" She replied, "I cry a lot lately." And I responded, "I can't cry at all, and that's worse." It was a totally spontaneous exchange, and afterward I thought of what an unusual conversation that had been. Did I really say that to a woman I had just met? During a later hospitalization on a rehabilitation floor, I was placed in a room with three women. This departure from customary hospital procedure was necessitated by crowding and scheduling problems, but it did not bother us occupants. None of us could get out of our beds, and an attempted molestation by me, or by one of the women, would have been hailed as a miracle.

As liminal people, the disabled confront each other as whole individuals, unseparated by social distinctions, and often they make strikingly frank revelations to each other. I have had informal conversations with paraplegic women that gravitated without direction to bowel and bladder problems. And one multiple sclerosis researcher, who has the disease herself, says that male informants usually tell her voluntarily about their impotency problems. This openness facilitates fieldwork among the disabled by disabled researchers, although one must be careful not to assume that his own experience of impairment is the same as his informant's. Another result of this democracy of the disabled is that, after they have gotten over their initial aversion—which only adds to their isolation—many seek each other's company, often through membership in disability organizations. There they find fellowship and a refuge from a world that commonly relegates them to its margins.

The disabled person fits into the mold of liminality far better than into the model of social deviance followed by sociologists. Writing about ritual process in primitive societies, Victor Turner says, ". . . liminality is frequently likened to death, to being in the womb, to invisibility, to darkness, to bisexuality, to the wilderness, and to an eclipse of the sun or moon."23 How well this fits everything we have discussed: the occasional rumor of my death, the social invisibility of the disabled, the attribution of asexuality in the popular mind, the unisex hospital room, and the blurring of sex roles within the community of the handicapped. The disabled are more than deviants. They are the antiphony of everyday life.

Just as the bodies of the disabled are permanently impaired, so also is their standing as members of society. The lasting indeterminacy of their state of being produces a similar lack of definition of their social roles, which are in any event superseded and obscured by submersion of their identities. Their persons are regarded as contaminated; eyes are averted and people take care not to approach wheelchairs too closely. My colleague Jessica Scheer refers to wheelchairs as "portable seclusion huts," for they are indeed isolation chambers of a sort.24 So, too, are the dwellings of the handicapped, and my associate Richard Mack reports on the plight of many poor, black paraplegics who live in New York City walk-up apartments that they can leave only when they are carried down several flights of stairs—which means seldom.25 They are prisoners.

The disabled in America are pulled back into themselves by

their own sense of loss and inadequacy, an impulse to withdraw that conspires with their devaluation by society to push them further into isolation. Add to this the fact that they confront a physical environment built by people with whole bodies for people with whole bodies, and one may well wonder how any of them manage to break out into the world. But they do, and in increasing numbers.

6

THE STRUGGLE FOR AUTONOMY

To live means to keep on existing. Every day is a victory . . . a victory felt as a triumph for life.

—Frantz Fanon,
The Wretched of the Earth

A friend once remarked that there seem to be far more people in wheelchairs these days than in the past, a casual observation that was surprisingly accurate. Back in the dark ages before antibiotics, people with damaged spinal cords commonly died from infections. Pneumonia found easy prey in persons whose weakened chest muscles and diminished lung capacity rendered them unable to cough up fluids, and recurrent urinary-tract infections eventually caused renal failure. It is the development of antibiotics over the past forty years that accounts for the explosion of the wheelchair population. In 1935, I would have died within a few years after becoming paraplegic. There are indeed more of us, if only because we stay alive longer.

With the help of modern technology, hospitals are able to keep alive babies born during the second trimester of pregnancy, although many survive with brain damage and other deficiencies. "Preemies" and normal-term children born with birth defects have become the center of political controversy in recent years. So-called right-to-life groups, in complicity with politicians who curry their favor, have intervened to prevent doctors and hospitals from withdrawing life-support measures and letting disabled neonates die. This was the issue in the "Baby Jane Doe" case, which involved proposed surgery on a baby born with spina bifida and Down's syndrome. How the baby's family was supposed to pay the huge and continuing medical expenses was another matter, for the same administration that intervened on the infant's "behalf" had gutted some of the programs that would have helped pay for her care. The matter was settled in June 1986, when the Supreme Court overruled the government position and left such decisions to the discretion of the parents and their doctors.

Whatever the costs, American values clearly are on the side of maintaining and prolonging life, and few people stop to reflect on its quality. This is to be expected in a culture such as ours, which seeks eternal youth and doubts the immortality of the soul. And it is also predictable that, given those cultural peculiarities discussed in chapter 5, the same society that sustains the body beyond its normal limits shies away from the results. The very old are avoided or treated as infantile, and the disabled are pensioned off and relegated to the status of outsiders. But at least they are alive, where once they would have been dead, and whether this is good or bad is irrelevant. The problem of the disabled person is very simple: You are just as alive as you always were, and what are you going to do about it? The disabled have the luxury—or agony—of choosing whether or not to take part in life. And they, more than others, know that if their decision is affirmative, they face a long and bitter struggle. The other reason for my friend's observation about the increasing prevalence of wheelchairs is that more and more disabled people are rising to meet the challenge to move out into the world.

There are about a million Americans in wheelchairs today, and their pathways to that condition are many, but whatever the reasons for their debility, their ecological problems are much the same. Discrimination against them crosses the boundaries of etiology, and the wheelchair itself has limitations no matter what the cause. And just as there is a pattern in discrimination against the handicapped, so also is there one in the epidemiology of motor disability, for it does not occur randomly in all sectors of the population. Injury to the spinal cord through accident or mishap is one of the most common causes of paralysis, and it has occurred disproportionately among young, working-class men. This is due in good part to the fact that wartime wounds— a bullet or a piece of shrapnel that nicked or severed the spinal cord—account for a large number of wheelchair users. Not only were the victims almost invariably men, but most of those who fought in Vietnam were drawn from the lower class. I doubt whether the American public fully appreciates the class bias of the draft during that era, but it was striking to everyone in higher education, one of the major routes at that time for draft deferment of the privileged. And college students were reasonably safe after graduation, for I cannot think of a single Columbia alumnus who was drafted. I am sure some were, but I am confident also that few served in the infantry. That was the domain of blacks and blue-collar whites, who thereby became the highest risk group for spinal cord injury.

The same class and gender factors occur also in civilian life. The chief causes of traumatic paraplegia and quadriplegia in American cities today are gunshot wounds, and, not surprisingly, most of the victims are lower-class males; many are also black, for this kind of thing happens much more frequently in Harlem or Watts than on Manhattan's Upper East Side or in Beverly Hills. Contact sports are another cause of paralytic injury, and here again the principal players are usually young, working-class men. Still others are injured on the job, continuing the same pattern. Automobile accidents are yet another source of trauma, and, as every insurance company actuary knows, they are correlated closely with young men and drinking. On the other hand, multiple sclerosis befalls women more commonly than men, although not in sufficient numbers to offset the male majority in traumatic cases. Many other sources of motor impairment are neutral to sex and class, although the sharp reduction in earning potential experienced by most wheel-chair users often relegates them to lower-class status, no matter where they started.

Disabled people have a great deal in common, but what they share most is an inimical environment—human and physical barriers so great as to drive many into isolation. Those who fight back, who assert themselves, who transcend those hurdles, do so out of folly or courage—which are in fact the same thing.

To understand the struggle of the disabled to win their rightful place in society, we must first consider the laws governing the battle, for they have many legal weapons. The Magna Carta of the lame, the halt, and the blind is the Rehabilitation Act of 1973, passed over the veto of Richard Nixon, who thought it would be too costly. Modeled on the Civil Rights Act of 1964, it extends the protection of the Fourteenth Amendment to the handicapped. Its key provision is Section 504, which stipulates that any facility or activity funded in whole or in part by the United States government must be made available and accessible to the disabled. This obviously covers post offices, federal courts, and other government facilities, but it also extends down to the state, county, and municipal levels. Almost every school and library district in the country receives federal support, as do many recreation facilities, hospitals, and other services. Public transit around the nation, from New York's gigantic MTA to the bus line in Sioux City, Iowa, is subsidized by federal funds, and most of them are still inaccessible. The applicability of Section 504 is not limited to public institutions and programs, for private universities and colleges rely heavily on the federal exchequer for support of research, scholarships, and construction. The Rehabilitation Act cuts a wide swath.

Section 504 has had generally salutary results, but the legislation did not include regulations for its enforcement, leaving that in the domain of the executive branch. A set of rules was drawn up by the Carter administration and promptly put on a shelf by the new administration in 1981. Since then, we have had either no regulations or weak ones. Despite this gap between the spirit of the law and its implementation, the wording of Section 504 has set an ideal of nondiscrimination backed by the full moral authority of the Congress.

Another milestone in handicapped civil rights was Public Law 94–142, the Education for All Handicapped Children Act of 1975, which mandated mainstream education for disabled children, or its equivalent in special programs. Wherever possible, handicapped pupils must be admitted to public schools and be allowed to attend classes with able-bodied students. If some special problem, such as a learning disability, makes this impossible, then alternate facilities have to be provided at public expense. This act dragged school boards and administrators kicking and screaming into the kinds of facilities that had been long established in Europe. One of the reasons for their resistance was that, like Section 504, it mandated changes but provided no funds to implement them. Finally, state governments across the country have passed a variety of measures barring discrimination against the disabled in private and public employment, housing, and places of public accommodation. It is to these laws that we owe accessible privately owned establishments, such as theaters and stores.

Let us begin a survey of the world of the motor-disabled with a consideration of the physical environment and its obstacles. First of all, there is no way that a wheelchair can climb stairs; one has to be carried up or down or bumped one step at a time, and this can be a tricky business. I have been dropped by inexperienced people, and I now routinely refuse invitations to houses with more than two or three steps. This problem places severe limitations on choice of housing. It also imposes a sharp restriction on territorial range, for one cannot go to work or seek relaxation in places with steps, nor can one climb bus steps or subway stairs.

The outside world is full of pitfalls. The best place for a wheelchair is the suburbs, because the streets usually have sidewalks, and, even where there are no curb cuts or ramps, at least it is possible to cross streets by using driveways. Rural districts, with narrow roads and no sidewalks, are bad wheelchair country; it's easy to be sideswiped. Paths can be used, of course, but most wheelchairs have thin, solid tires and no springs, making for bone-rattling travel on unpaved surfaces. As for cities, maneuverability is contingent on whether the curbs at street corners have been cut and ramped. The cost of doing this throughout an entire urban area is mind-boggling, and many cities follow New York City's policy of putting in curb cuts only when they repair a sidewalk or street. As a result, sometimes one can get from sidewalk to street via a curb cut but is unable to get over the opposite, uncut curb. Paraplegics with good upper body strength can get over curbs by pulling backward sharply on their wheels, making the front wheels rise several inches off the ground, and then pushing forward hard in a single motion that carries the front wheels over the curb and the back wheels up it. Needless to say, quadriplegics can't do this, because their arms are too weak. They either must be accompanied or depend on the kindness of strangers.

The problems posed by the streets, however, are relatively minor. In every meeting of the handicapped I have attended, the major complaints were about lack of accessible housing, employment, and transportation. Housing difficulties vary from neighborhood to neighborhood, city to city, and region to region in a gradient that goes from bad to impossible. In New York City, which has an apartment vacancy rate of under three percent, the situation is almost impossible, even for able-bodied people; the vacancy rate in Manhattan is down to one percent, and most of the few unrented units are being kept off the market by landlords planning to convert the buildings to cooperatives. Considering that much of the available housing also is inaccessible for wheelchairs, the pool shrinks rapidly. As if this were not bad enough, the rents charged for New York apartments are astronomical. Non–rent-controlled studio apartments in 1986 rent for $800 to $1,000 a month, and one-bedroom units are more, placing them far beyond the range of most people, not to mention the handicapped. The housing situation in other parts of the United States may be less severe, but cost and availability are everywhere a problem for the disabled. There is a critical nationwide shortage of low-cost housing, a lack so serious that homelessness is higher now than during the Depression—and it shows every sign of becoming worse. The disabled face a bleak situation.

Not all apartments are usable by the motor-disabled. The building ideally should have no front steps, although most wheelchair users can handle one or two with a bit of help. And, if the individual is to have any freedom of movement, it is essential that the building have an elevator. Unfortunately, much of the low-rent private housing in urban America is in walk-up apartment houses, making the plight of the disabled all the more acute. And even if a building is accessible, the apartments may not be. Doors must be wide enough to let a wheelchair through, and halls must be broad enough so that a wheelchair can make a right-angle turn into a room. To be usable, most bathrooms require modification, if only the installation of grabrails next to the toilet and the tub. And kitchens must be designed to bring facilities and utensils within reach. Landlords are loath to make alterations, and most, in any event, are averse to having handicapped people in their buildings. "It's just been rented," a phrase long used to deny housing to blacks in this country, is now a vehicle for keeping out the handicapped. Landlords rationalize this kind of discrimination by citing fear that the disabled are prone to accidents and will make insurance premiums go up, but underlying this is an aversion to anybody who is different. It is fruitless to try to appeal to the consciences of landlords, or at least New York ones, for they generally are a cold-hearted lot. It is not sympathy that the disabled want from such people; it is simple equity. And this is sought increasingly through legal means.

The disabled person's best chance for an apartment lies in public housing, and many municipalities give special preference to the handicapped. Moreover, ten percent of federally supported senior-citizen housing must, by statute, be adapted to and reserved for the handicapped. But this often places people in their twenties among neighbors in their seventies, and the elders commonly frown on the behavior of the young in their midst. They play loud music, hold noisy parties, and are even suspected of indulging in sex. The elderly, too, believe that the disabled should not be enjoying life all that much. Although subsidized public housing has been one of the best sources of apartments for the disabled, recent federal housing policies, coupled with enormous deficits spawned by the tax cuts of 1981 and several years of military buildup, have effectively terminated that program. Beyond the provision of housing, many state and local ordinances now prohibit discrimination against the handicapped in sales or rentals, a logical extension of the ban on racial discrimination, but all these laws are massively breached every day by real estate interests.

A disabled person may be able to get in and out of his dwelling place only to find himself unable to go anywhere due to lack of transportation. This is a major problem for all disabled people, especially the wheelchair-bound, and it has been the subject of vigorous politicking by the disabled. This is curiously reminiscent of the black civil rights movement, which had its roots in Rosa Parks's refusal to sit in the rear of a bus in Montgomery, Alabama. The protest by the disabled, however, is less an effort to overturn unjust laws than a fight to ensure the enforcement of a just one, Section 504.

When considering the dilemma of the disabled, it is important to emphasize that until the Rehabilitation Act of 1973, not a single thing was done to meet their transportation needs, except for van or ambulance transportation to hospitals and clinics—a free service for Medicare or Medicaid recipients but a prohibitively expensive means of travel for those not receiving public support. There was not a single bus, subway, or rail system in this nation usable by a wheelchair-bound person, despite the fact that the disabled helped to pay for these facilities with their taxes. And up to the early 1970s, the country's airlines refused routinely to carry the motor-impaired, on the specious grounds that they could not be evacuated in an emergency and could even impede the escape of others. One of the first victories of the disabled movement was in forcing the airlines to abandon these policies. The ease of the victory, however, was due in large part to the fact that the disabled could be accommodated at negligible cost to the airlines, which discovered that, after all was said and done, they had gained new paying passengers.

Surface transportation has been a far more intractable matter, for vehicle adaptation is very expensive. In order to accommodate wheelchairs, elevators would have to be installed from street to platform in all subway and elevated stations; places for wheelchairs and tie-downs to keep them from rolling would have to be provided in every car; and something would have to be done about the wide gaps between platforms and cars, a short step for most people but a dangerous abyss for the small front wheels of a chair. Recently built subway systems, such as Washington's Metro or San Francisco's BART, were built with handicapped accessibility in mind, but those of New York, Chicago, Philadelphia, and Boston were not. Because of the huge expense involved, there has been great resistance in all these cities to large-scale renovation of their old mass-transit systems.

Most of the large cities have resorted to a provision of Section 504 that accepts substitution of other facilities. Some have experimented with minibuses with wheelchair lifts and others have tried taxicab vouchers, on the grounds that it is far cheaper per ride than the cost of reconstruction. This sounds good, except that taxis routinely drive past people in wheelchairs. In New York, a black person has a better chance of catching a cab to Harlem than somebody in a chair has of going anywhere.

The most common means of providing accessible transport has been the wheelchair lift-equipped bus, but their adoption by New York City is a cautionary tale. Richard Mack reports that it was only after strong pressure by disabled activists, backed by sympathetic elected officials, that the Metropolitan Transit Authority consented to buy new buses outfitted with lifts.[1] And in the standard bureaucratic manner, they bought GM buses, which have the lift in the back door, and the ill-fated Grumman Flxibles, whose lifts are in the front door. There then followed months of "driver training" to operate these simple devices, after which there were further delays in implementing routes with accessible buses. The MTA clearly was stalling, and the disabled turned out in protest. One doughty lady shifted from her wheelchair to the steps of a bus that refused to take her, and she remained there for seven hours while a squad of policemen and embarrassed MTA officials pondered what to do about her. The episode made the evening news on every TV station in New York, and the MTA knew the game was over.

When service finally did begin, Mack reports, drivers still passed up disabled people, claiming either that the lift was broken or they didn't have a key. Despite the unwillingness of many drivers to operate the lifts, more and more lift buses were put on an increasing number of routes—until the inevitable New York calamity occurred. After a series of vehicle breakdowns, the MTA withdrew all its Grummans from service, leaving the system with less than half of its lift-equipped fleet. Those left were the GMs, in which the driver has to go to the rear of the bus to run the lift. This did nothing for the dispositions of the drivers— not a cheery lot under the best of circumstances—and on the very day I was writing these lines, I read in The New York Times of an elderly lady in a wheelchair who said tearfully that she no longer goes out because she can't stand the surliness of the drivers. Not all the news has been bad, however, for in 1984 New York Governor Mario Cuomo entered into an agreement with the city and the MTA in which elevators would be installed in fifty subway stations scheduled for modernization, with feeder bus and van lines connected to them. After years of trying to use federal statutes, the disability groups won their case by resort to the state building code, which requires accessibility for the disabled in any new or newly renovated place of public accommodation.

The best way for the handicapped to get about, of course, is by private car. This poses difficulties in many cities because of high insurance rates, vandalism, the difficulty of parking, and theft. Besides, most people cannot afford a car on a Social Security stipend. But for those who have the money, a variety of driving aids are available. Ordinary sedans can be fitted with special power-steering equipment and hand brakes and throttles sensitive to light pressure. One man whose polio left his arms useless steers with his legs. Quadriplegics whose upper bodies and arms are too weak to transfer from chair to car seat usually buy specially equipped vans. These come with wheelchair lifts, controls, electrically operated doors, and a host of gadgets allowing them to drive. But a fully outfitted van costs well over $25,000! And the thousands of people too disabled to drive any vehicle must rely on a family member or paid aide to chauffeur them. The travel problems of the disabled are a long way from being resolved.

State laws commonly protect the handicapped from job discrimination, but bias is even more difficult to prove here than in housing. Employers claim that hiring the disabled would jeopardize their workmen's compensation and health insurance contracts or leave them liable for suits—all of which are untrue. Closely related is a belief that the physical ailments of the handicapped will result in excessive absences, a canard that is reminiscent of the old chestnut that female workers will be absent three or four days every month (a weakness that curiously only keeps women from good jobs). Some employers claim that disabled workers are not as productive as the able-bodied. Actually, the reverse is true, and many disabled people compensate for their physical deficiencies by becoming overachievers. Another excuse used frequently is that a disabled worker might repel customers or other employees, a possible problem that can be neutralized effectively by strategic placement of the person. Besides, it has been the experience of most observers that people eventually accept, then routinize, the presence of the disabled.

Employment discrimination is nonetheless prevalent, and in places where it would be least expected. I met one young woman wheelchair user who is a graduate of an Ivy League college and one of the country's foremost law schools. Her credentials were so good that she had no trouble obtaining interviews at leading New York law firms and corporations, but, mysteriously, none resulted in a job offer. Finally, after a year of searching in New York City, the world center of the legal trade, she landed a post with a branch of government; she was the last member of her law school class to find employment. Was there discrimination? Of course there was! Did her prospective employers know the law? Forward and backward. Could this discrimination be proved in a court of law? Not in a million years—these people are too clever to be caught. And it is against this kind of prejudice that the handicapped must flail every day.

The disabled persons' own physical capabilities and deficiencies are important factors in their employability. Almost all forms of heavy manual labor are beyond the reach of wheelchair users, although I know several skilled, paraplegic mechanics who do excellent light assembly and repair work at benches. Actually, paraplegics whose hands and upper body are intact are perfectly capable of doing these kinds of tasks. Recognizing this, many vocational rehabilitation offices used to have a policy of sending their clients to watchmaking school, which became the paraplegic's equivalent of broom-making and chair-caning among the blind. Unfortunately, they continued doing this long after the introduction of the cheap digital watch, and many people found their skills obsolete before they ever had a chance to practice them. And to top the matter, even if a person gets the needed training, and even if he fights prejudice successfully and finds a job, the workplace may contain physical barriers that make it inaccessible to wheelchairs.

The employment picture for quadriplegics is even more of a problem, for most lack the upper body strength and manual dexterity needed for mechanical work. And it is the policy of most vocational rehabilitation agencies to retrain only those people whom they deem employable. They often rule out quadriplegics on these grounds, in effect telling them to collect Social

Security disability and remain quietly out of sight. Needless to say, there is high dissatisfaction with state rehabilitation agencies, a feeling that has been intensified since 1982 by inadequate federal support for vocational rehabilitation. The logic of these "economies" is elusive, for the handicapped rehabilitation program is very cost-effective.

Reeducation of the disabled converts a person from tax recipient to taxpayer, and, over a lifetime, this can return hundreds of thousands of dollars in savings to the public coffers—many times more than the modest costs of his education. Economists have placed the rate of return on investment in vocational rehabilitation at nine to one, a figure that startled me when I first read it, but apparently it is a modest estimate. In his admirable 1980 book, *Rehabilitating America*, Frank Bowe notes that the handicapped are capable of full economic independence to an extent most people would think impossible.[2] To cite one example, Israel, which is labor-short and unable to afford an army of pensioners, has returned ninety-seven percent of its war-disabled to economic productivity, providing incalculable benefits for the country's economy as well as for the wounded veterans.

Bowe writes that the increasing burden on the taxpayer for support of the old and the disabled could be eased greatly by a sensible policy of job training and placement, with annual savings in the tens of billions of dollars. Suppose, to give the most expensive example, that federal funds were used to send a disabled person through an Ivy League college at an annual cost in tuition and maintenance of $20,000. If we throw in the purchase of a van at the end of the four years, the total outlay comes to $100,000, an amount that would leave most legislators in shock. But look at the alternatives. If the same handicapped individual were to be supported by Social Security and Medicare, the annual cost to the taxpayer would be in the neighborhood of $7,000, a rather low estimate. The employed college graduate, on the other hand, would cost nothing, and could well be returning to the U.S. Treasury income and Social Security taxes of

$7,000 per year, another conservative figure. At this rate, the federal investment would be paid off in only seven or eight years, after which it all would be clear profit. Despite the obvious wisdom of rehabilitation, it has been recent government policy to economize by reducing rehabilitation budgets and dropping people from disability rolls. It is a nickel-and-dime approach that wastes both public funds and the lives of individuals.

One of the more encouraging developments in rehabilitation in recent years has been the rapid growth of the service economy. The new economy and its technology have been a liberating force for the handicapped, for their infirmities do not inhibit them from working in front of a computer terminal, interviewing welfare clients, or serving as corporate financial analysts. Indeed, they even can take part in primary production, for in an age of robotics, the controllers of the robots need education, not brawn. These are new horizons for the disabled (and for women, too), and the age has just begun. The American labor market has now started to draw on the 1.7-children-per-woman pool of new job entrants, and many industries already are having difficulty recruiting young workers. I suspect that disabled people suddenly will begin to look much better to employers in a future seller's labor market.

The effects of the new technology on the disabled are dramatically evident in the use of computers, which have replaced typewriters, calculators, filing cabinets, paper, drawing boards, and a host of other things that are intractable for the handicapped. Computers open doors, turn on a stereo, read to the blind, and soon will be voice-activated. I could not have written this book without a computer, but it still boggles my mind that I may soon dictate a book into one end of a machine and get a typed manuscript out of the other end. Many disabled people have gone into a career of computer programming, and even more have entered fields that would have been closed to them before. Most of the handicapped are capable of using computers, and I know one well-paid, quadriplegic programmer who holds a stylus between his teeth. Computer uses are too numerous to list, and there already is a bibliography of dozens of books and pamphlets on computers for the disabled.

One does not enter this new economy by attending tinker school for six months; one enters it via higher education. Schools across the country have experienced a sharp surge in handicapped enrollments, despite the fact that financially straitened vocational rehabilitation offices tend to make fewer referrals. Among the handicapped, the two most popular institutions in the New York City area are Hofstra University on Long Island and Ramapo State College in New Jersey. Every part of these campuses is accessible to wheelchairs; they have installed ramps and electrically operated sliding doors; and they have modified dormitory rooms, bathrooms, and elevators. It is encouraging to report that fellow students and faculty at those schools at first were taken aback by the handicapped invasion, but their presence became routine after a few semesters. The disabled no longer stand out from a diffuse able-bodied-student background, their centrality in the scene bespoken by averted eyes. They have now faded into that background.

Most universities and colleges have made alterations to conform to the spirit, if not the letter, of Section 504, a task of some magnitude on older campuses. Columbia is a fine example of blocked accessibility, for its location on the crest of Morningside Heights divides the campus into four distinct levels connected to one another by steps. Most of the buildings date from the first decade of this century (Columbia was founded in 1754, but it has occupied its present site for only about ninety years), and their handsome outlines, drafted by the architects McKim, Mead, and White, are among the university's prides. But every one of these beautiful buildings is approached by two to four granite steps, and inside the lobby are a few more made of marble. Not surprisingly, very few students in wheelchairs have ever enrolled at Columbia.

When I returned to the campus in a wheelchair in 1977, I was in a new environment, for the old familiar setting had become impossible to navigate. There were a few metal ramps here and there, some so steep that I needed two helpers to go up or down, and all were a slippery menace in rainy weather. The anthropology department had to assign an assistant to get books for me and ferry me around. In exasperation, I wrote a letter to William McGill, then president of Columbia, and he came to my office as soon as he received it. He proposed setting up a committee under my chairmanship to tackle the problem and advise him on what to do. This sounded suspiciously like the old bureaucratic game of putting complainers on study committees, but McGill reassured me that the university was seriously concerned with compliance with the federal statute and had already commissioned an architectural survey. Serving on such a committee, let alone chairing it, was the furthest thing from my mind, but he had me fair and square. I accepted. My confidence was shaken a bit when I called the Office of Equal Opportunity and Affirmative Action, whose director was to be executive secretary of the committee, and heard an answering machine, which instructed me to call a non-working number. A new director, Rosalind Fink, arrived a couple of weeks later, however, and soon became the driving force behind the committee's work.

The university kept its word, and McGill's support of the committee was continued by his successor, Michael Sovern. Over a five-year period, the university spent more than $700,000 on alterations that we recommended. An elevator was installed to lift people from one campus level to another, and permanent, low-gradient ramps interconnected other levels. Building entrances were ramped, and interior steps were bridged by wheelchair lifts. One such lift made the gymnasium accessible; even if the motor-disabled can't use the squash courts and running track, they can at least enjoy the weight rooms and the swimming pool. The modifications didn't stop there. We renovated restrooms in most of the buildings to accommodate wheelchairs, and we lowered elevator control panels and included Braille floor numbers. We even lowered many drinking fountains. After five years, there are only two buildings that remain inaccessible, and this problem is handled by moving a class to another building if a wheelchair-bound student wishes to attend. It was a long and complex task, but in all my years of serving on university committees, this is the only one that has given me any satisfaction. We were given decision-making power and fiscal responsibility, and we held costs at one-third of the original estimates.

Most campuses are now open to disabled students, and, in a period of declining enrollments, schools are glad to have them. These students are not training just for computer work; they have enrolled in medicine and law, and they are undergraduate majors or graduate students in almost every discipline. Their numbers are not great, but they are growing every year. They will suffer discrimination in the job market, and they realize that, but they also know that in the kinds of careers for which they are training, their infirmities are largely irrelevant to their potential.

Many disabled people enter professions that offer social services to the handicapped. They become psychologists, social workers, "peer counselors," speech therapists, job consultants, and so forth. These are fields in which their frailty is a strength, and they excel; they are also able to make good livings. Surprisingly, the disabled sometimes encounter discrimination even in these jobs, but far less than in the private sector; these are pathways to autonomy. And this group is interesting for another reason: They are very active in disability advocacy organizations.

The Rehabilitation Act of 1973 did much to galvanize the consciousness of disabled people and make them aware that they are more than a motley assortment of damaged individuals: They are a class. The birth of ameliorative legislation provided a weapon in their fight for civil liberties, and, equally important but usually overlooked, it publicly ratified the feelings of many that they were the victims of systematic discrimination. They are an oppressed people. As such, the disabled have a common collective interest, and the fate of all is involved in the fortunes of each. Out of this new self-awareness arose a plethora of organizations of and for the disabled and a new spirit of activism and militancy. The historian Crane Brinton found that revolutions do not arise from the depths of despair, but rather during times of heightened expectations, periods when the desirable is seen to be attainable. The Rehabilitation Act of 1973 and its Section 504 ignited that hope among the disabled.

There were, of course, many disability advocacy groups before the 1970s. Most, however, were oriented toward providing medical support, as in the case of almost all the ailment-specific charities, and none of them were run by the people being helped. Perhaps the oldest and most prominent of the politically oriented self-help groups is the Paralyzed Veterans of America (PVA), which was formed immediately after World War II by veterans with spinal cord injuries. The PVA started its life by fighting the Veterans Administration for the rights of its members and lobbying Congress on their behalf. They have since expanded their horizons, and the New York office of the Eastern Paralyzed Veterans of America provided both material support and political and legal expertise in the successful campaign to open up New York City's subways. The PVA was thus a leader and a model for the blossoming of disability groups in recent years.

Another important incentive for the disabled rights movement's marked vigor has been recent federal policy. Since 1981, each of the budgets sent to Congress by the Executive branch has contained severe cuts in federal spending on programs vital to the disabled, and only their rejection in whole or part by Congress has kept a desperate situation from becoming a total disaster. I have already mentioned the demise of federal housing programs, once a major source of shelter, and the loss of funds for vocational rehabilitation. Deep cuts in medical support and food stamps have also compounded the wounds of the handicapped. Most of the wheelchair brigade are unemployed and supported by Supplemental Security Income (SSI) or Social Security Disability Insurance (SSDI). In 1986, SSI paid $367 per month, and the average stipend of SSDI was about $870 for persons with two dependents. Food stamps help them eke out the difference between these meager stipends and a survival income. Social Security also provides inclusion on the Medicare rolls, but those benefits also have been reduced in the last four years.

According to the 1980 census, there are 12,320,000 Americans between the ages of sixteen and sixty-four—some 8.5 percent of the total labor force—whose physical impairments limit or totally prevent gainful employment.[3] They are, on the average, some sixteen years older than able-bodied workers, and significantly less educated. Frank Bowe has interpreted this wisely as a cause of disability rather than an effect, because it is the less educated who do manual labor and serve in infantry platoons, all of which are high-risk occupations.[4] As might be expected, the income levels of the disabled population are far below those of the nondisabled. The median income of the handicapped was about sixty-two percent of the nondisabled in 1980.[5] And whereas one out of every ten able-bodied people fell below the poverty level in 1980, the figure for the disabled was one in four. This disparity is due in part to educational deficiencies, but it stems mostly from the fact that 58.2 percent of disabled men and 76.5 percent of disabled women are totally out of the labor force, and most depend heavily on public assistance.[6]

The reductions in benefits since 1981 both frightened and aroused the disabled, and many joined disability organizations as a way of making a collective protest. Their alarm turned in many cases to panic over an even-greater threat. In 1980, Congress ordered the Social Security Administration to review its disability rolls and drop those whose conditions had improved enough to allow them to return to work. The administration has, since 1981, used this call for routine reassessment to conduct a wholesale reduction in the program. Claim examiners were given a set of harshly restrictive guidelines, and they had to meet quotas for dropping people. Hundreds of thousands of termination notices were sent out after clerks made cursory examinations of files; there were few medical reevaluations. The notices advised recipients of the procedure for appealing the decision to administrative judges, who also had quotas to meet. Due to the anticipated flood of appeals, this process could take months; if that failed, one could appeal in a federal district court, which could take years. And all the while, the hapless victim would be without income.

The havoc caused by this program was great. I remember one former psychiatric social worker, whose worsening multiple sclerosis had forced her into retirement, asking me tearfully what she could do. I had no answer other than to suggest she appeal and, in the meantime, apply for welfare. Thousands had to take this route, which meant one more humiliation heaped on them by an uncaring government. Beyond the shame many felt at being on relief, local welfare usually pays much less than Social Security, despite the self-serving redneck myths of welfare queens and relief luxury. Many of the terminated people also had to go on welfare in order to obtain Medicaid coverage, for their Medicare eligibility was also canceled at the same time as their disability pensions.

The impact on the disabled was nowhere greater than among the mentally ill, who often failed to understand what had happened to them; many just threw away the letters of termination. These cases were reported widely in the media, and the government eventually was forced to announce plans to loosen the guidelines. The policy changed very little, however, until a moratorium on reviews was forced by Congress. A very large percentage of those who appealed their cases to the federal courts, which brook no quotas, were reinstated, but the sheer weight of numbers overwhelmed the resources of legal assistance groups, which also had suffered a reduction in federal funding. To make matters worse, the Department of Health and Human Services still refused to treat court decisions in favor of the disabled as class actions applicable to other cases of the same type, despite court orders to do so. This forced each appellant to press his own separate case, a procedure that has clogged court calendars around the country. A sick or disabled person could die before his case was heard—and hundreds have. The latest maneuver in this war against the disability program was a 1986 proposal that special courts be established for these appeals, with the judges to be nominated by the president, of course. It is just this kind of dogged, insensate persistence that has made many disabled people feel that they have been abandoned by their own government.

The effect of all this was to swell the ranks of the disability groups and foster militant political action. There were rallies, public open-air demonstrations, lobbying of members of Congress, letter-writing campaigns, picket lines, and even a few sit-ins. They had good results. Legislators listened sympathetically and were impressed by the sizable crowds drawn by the disability organizations. This, in itself, was a major feat, for the transportation problems of the disabled made it difficult for them to get out to protest these very problems—a neat kind of double bind. The volume of mail from the disabled alarmed even the most conservative elected officials, and support for the administration waned on such issues.

As the worst of the onslaught subsided, many groups lost some of the vigor and enthusiasm bred by good and just battle, but the struggle had produced significant achievements. It was, in the first place, an education in political organizing and tactics, and in laws regarding the disabled. But, most important, it yielded a measure of victory, and the membership discovered that it is possible to fight back and win; they did not have to be the passive recipients of everything done either for them or to them. This social service–client attitude is pervasive among many disabled people, whose dependence on outside agencies and people becomes internalized in character structure and expressed in worldview. They often ask representatives of disability organizations, "What can this group do for me?" The usual answer is, "Nothing. It only gives you a way to help yourself." And they did. Many took courage in this involvement, for, in a way, they were no longer quite as disabled.

The most lasting benefits of any struggle against perceived oppression are not the tangible gains but the transformations of consciousness of the combatants. The disabled renewed and repaired their damaged egos, and they saw themselves even more clearly as a common interest group, with shared goals. They also found fellowship in the organizations—other people who had similar problems and to whom they could reach out in that peculiar egalitarianism that prevails in social liminality. A person in a wheelchair need not be concerned about how another individual in a wheelchair will react to him. He doesn't have to watch for the shifting glance and the affected greeting, nor does he have to take pains to put the other at ease. It may be hard at first for a person to accept his or her common identity with the blind, the deaf, and the crippled, but once this barrier is broken, he or she finds a haven among them from the strained and flawed relations of the outer world. The disability movement is one of the best available forms of rehabilitation.

Another accomplishment of the handicapped organizations has been to force awareness of their presence and needs on the public, a goal also of the 1981 International Year of the Disabled Person. The best way to fight prejudice is to jam the issue down people's throats by intruding upon their perspectives and perceptions, by bringing the despised or shunned into unavoidable contiguity and association with the world. One means of doing this is by law. Contrary to the folk adage, you can indeed legislate morality, and we do it all the time. The social psychologist Leon Festinger wrote in his book *A Theory of Cognitive Dissonance* that when values and attitudes are confronted by a contradictory and inescapable social reality, people often change their views.[7] In a less jargon-laden time, we used to say, "If you can't beat 'em, join 'em." Southern whites now routinely accept blacks in the sections of buses and theaters that once were their exclusive domain, and I saw an entire suburban block liberalize its racial attitudes after the sale of a house to a black family became a *fait accompli*. And nobody looks twice anymore at a policewoman. In similar fashion, as the able-bodied population becomes more accustomed to the presence of the disabled in their buses, their theaters, their schools, and their workplaces, their prejudices surely will become muted. I believe a start already has been made in that direction.

The major goal of the handicapped movement is not to foster dependency, but to move the disabled into the mainstream of society as autonomous individuals. As we have seen, the obstacles are formidable. The handicapped individual first must fight all his own impulses to flee back inside himself, to wrap himself in the mantle of his impairment. He must overcome a sense of inferiority, he must break out of isolation, he must reach out to a world that does not welcome him. Our society erects walls of discrimination and inaccessibility, both physical and social, but the handicapped keep trying to scale those barriers, to break out of themselves, to escape the web of constraint, to force their way into a full and rounded life.

In their search for autonomy, many of the disabled have joined together in "independent living" groups. The goal of these organizations is to enable their members to live on their own as fully functioning participants in their communities. Whether one joins such an organization or does it on one's own, living autonomously requires certain physical capabilities. The individual must be able to get in and out of bed and on and off the toilet, and he must be able to feed and dress himself; for the motor-disabled, this requires considerable upper body function. People with weak arms cannot push downward to lift themselves from seats and then shift sideways to bed or toilet, the standard transfer move. And persons with stiff fingers cannot button shirts or tie shoes. Advanced quadriplegics thus require personal care, which may be supplied by a family member or an attendant paid either from private funds or by Medicare and Medicaid. And there's the rub. To be eligible for Medicare, one has to be receiving SSI or SSDI; to qualify for Medicaid, one must be virtually indigent.

Most disabled people would prefer to work, even at low-paying jobs, rather than live on the Social Security rolls, but if their earned income exceeds a very low limit, they lose both their pension and their government health insurance. This is a critical matter, as their medical expenses usually are much higher than those of ordinary people. Even if they are accepted by an employer's private plan, these generally do not pay for attendants and certain costly equipment. In this classic bind, many disabled people cannot afford to be employed!

The Social Security bureaucracy and its labyrinthine regulations often beget tragedy. One young California woman struggled to get a rather low-paying job, only to find that she had lost both her SSI stipend and Medicare. She could have accepted that, but then she was also informed that she owed the government $10,000 in overpayments, payable immediately, or else harsh measures would be taken. The woman committed suicide.[8] This is not an unusual case, although most of the victims don't take their own lives, for the disabled are constantly being forced out of the labor market by rules that allow no middle ground between total dependency and full independence.

In another case, a paraplegic New Jersey man organized and planned a project to build a rent-subsidized apartment complex for motor-disabled people who require only limited attendant care, which would be provided through the program by two resident employees. Presumably, this independence would allow the handicapped tenants to work. The man fought for the program through the Federal Housing Administration; through state, county, and local governments; through planning boards and zoning boards; and finally his dream came true. Construction started and applications for apartments, including his, were invited and sent to the FHA for processing. In due time, the FHA sent him a rejection notice, on the grounds that he made too much money in his county government job. The eligibility requirement for the project was so low that only nonemployed pensioners could qualify, which negated the entire purpose of the project—and in the process excluded its founder. It took two congressmen and both New Jersey senators to get the FHA to raise the limit. Once again, the efforts of the disabled to become autonomous, working citizens had been frustrated by the very agencies that were supposed to help them realize these aspirations. The system itself promotes dependency.

It takes a rare combination of intelligence, courage, and persistence to conquer the mental and physical quarantine thrown up around the disabled by a society that secretly sees in them its own epitaph. Yet young handicapped people are doing this every day, and they are winning. Despite all their troubles and struggles, the young attorney in a wheelchair now has an excellent job in her profession, and the county employee has moved out of his parents' house and lives in a comfortable apartment in the complex he originated. The intensity of purpose required by the drive for autonomy makes the successful people unusual.

They have entered the mainstream of social life, and they have done this through great determination and unflagging effort. But no matter how well they may become assimilated into society, their struggle sets them apart from their able-bodied fellows. They have a different history and follow a separate agenda; they remain part of the Other. Their otherness, however, is positive and creative, for their self-assertion is a profound celebration of life.

PART III

ON LIVING

7

THE DEEPENING
SILENCE

*Most things under the moon got slower
and slower and then stopped. . . . Soon his
body would be quiet, soon he would be
free.*

—Samuel Beckett,
Murphy

Six months after my spinal cord surgery, I began to notice that I was on a plateau, a very low one at that. My exercises walking back and forth across the living room showed no further improvement in either distance or endurance, and my semiweekly, one-hour physical therapy sessions seemed to be getting nowhere. During these home visits, the therapist put me through much the same routine as in the hospital rehabilitation department. I pushed my legs and arms against his resistance, got up on my hands and knees, rolled over, sat up, and transferred from bed to chair. I showed small increases in strength for the first six months, but that came to an end. Physiotherapy can do wonders, but only within the limits set by neurological conditions, and I had reached that outer boundary.

The signs of degeneration were unmistakable, small though they were. The process was so slow and differences were so

165

slight that they were not noticeable from day to day, or even week to week. I had ups and downs, however, and these were clues to the future. Conditions such as mine are subject to daily variation, a phenomenon that I attribute to the weather in lieu of a better explanation. It was on one of my bad days in 1978 that I discovered, when getting into the wheelchair, that I couldn't lift my right leg (the good one) high enough to place my foot on the footrest, a distance of only a few inches. Instead, I had to grab the leg above the knee and lift it onto the footplate, which is what I had been doing all along with the left leg. The next day, I was able again to lift the right foot, only to have the same setback occur the next week, then more frequently until the right leg couldn't be lifted at all. Occasional little failures, then, were harbingers of things to come.

The tumor was growing steadily, inexorably and glacially. A month-by-month chronicle of my neurological deterioration would be as interesting as watching grass grow, so I will summarize my physical regress from 1977 to 1979. After about a year, walking became more difficult. The left leg dragged even more heavily on the floor, and I could neither lift the right foot as high as before nor move the right leg as far forward. Finally, I was reduced to shuffling with both feet, and in late 1978 I gave up walking altogether for fear of falling. I could still get to my feet, however, and, with the support of a walker, travel a few steps to transfer from wheelchair to bed, armchair, or toilet.

There is a relationship of positive feedback between usage and capacity; that is, they amplify each other. The more one uses one's legs, the greater will be their capability; similarly, disuse leads rapidly to atrophy and loss of potential for function. As a result, the end of walking accelerated the deterioration of my legs, which ultimately sapped my ability to rise from a chair or bed to a standing position. As I described earlier, getting up was always a tricky business. Since the knees are flexed at a right angle when sitting, only the vigorous can rise using their legs alone. One sure sign of advancing years appears when a person has to push down on a chair seat or arm to get a much-needed lift-off boost. Similarly, those with infirm legs must rely on their arms to push them up to the point where their legs are straight enough to take over the rest of the job. My arms and hands had weakened slightly over the years, but in 1978 I still had ample strength for the initial push-up. The problem arose at the point where the legs were supposed to take over, a transfer of function that, unknown to me, had become increasingly delicate, operating within ever-narrowing tolerances. I had not realized how close my standing exercise had come to the limits of my waning leg strength until one day when I got halfway up and then fell back heavily into my chair. My legs had failed me.

My inability to get to my feet was the crossing of a Rubicon. The gradual, silent, insidious little increments of weakness suddenly had been translated into a major change of function. My legs were no longer strong enough to lift a 140-pound man— although they probably could still have levered a 120-pound one to a standing position—and my arms were not long enough or strong enough to push me higher. I knew that the ability to stand might return for a week or so, but the end clearly was near. It was indeed an ill omen, for on a cold, raw, and rainy November day in 1978, I tried to stand up to get in a car and couldn't. I remember the day well, for we were on our way to Margaret Mead's funeral.

Sitting in my wheelchair in the center aisle of Columbia's St. Paul's Chapel, where Margaret's Episcopalian service was held, I fell into a reverie about how deeply entangled in the university my life had become. The last time I had been in the chapel was in 1968, when an emergency meeting of faculty and administrators was held there to discuss the student strike. It was at this stormy meeting that members of my department criticized the university administration's handling of the crisis. Before that, Yolanda and I had been in St. Paul's in 1950, for this is where we were married. Both of us were Columbia graduate students in anthropology at the time; the campus is where our romance started and where it had its culmination.

My days at Columbia began in 1946, when I registered at Columbia College. I arrived on the campus seven months after being discharged from the Navy and a few days after the ship on which I had served as a civilian crewman had finished a run from Bremerhaven and Southampton. Higher education had always been an ultimate aspiration but a financial impossibility for me—and for millions of other products of the Depression. The GI Bill opened college education to us, and we entered the universities like a horde of hungry barbarians. For the first time in the history of this country, higher education became available to the excluded, the unwashed, the outsiders, the undesirables. College was no longer a middle-class monopoly, and with the breaching of the class barriers, ethnic and racial walls also began to crumble. Out of this came the most remarkable spurt of class mobility ever experienced anywhere, and the talents unleashed by this experiment produced an unparalleled and sustained growth of the American economy for the next three decades. In simple dollar terms, it was the best investment this country ever made.

Most of us came from families in which high school graduation was considered an achievement, and we had no idea what university life was like. I had little understanding of what a Ph.D. was, and none at all of how to obtain the degree. As for anthropology, the first time I ever heard of the subject was in 1947, when I asked a friend to recommend a course that would be easy yet interesting. "Anthropology," he replied immediately, and I said, "What in hell is that?" Like much of my generation, I had a wealth of experience in very unacademic pursuits, for we had seen sights and done things beyond the ken of most undergraduates. But we were intellectually unfinished and rough. Columbia took me in hand, formed me, brought out my potential, and encouraged me to become whatever I wanted to be, to do whatever my abilities permitted. It opened up a universe for me that was rich beyond anything I had ever imagined.

My undergraduate years in Columbia's still-vibrant general education program were a joy, a continuing feast. I discovered in philosophy the human quest for ultimate, or even contingent, meanings, and for an understanding of the human condition. In anthropology I found a perspective and method for probing these questions in an orderly, controlled way, a discipline whose magisterial range allowed one to study within its confines all things human. Columbia was the best place I had ever been, a better home by far than the one I had grown up in, better duty than anything I had known in the Navy. I had been in the service for three years, and I had worked in downtown Manhattan for a couple of years before that; it's hard to say which I disliked more. I became determined to stay in university life. It was in this way that the barbarians took over academia, which had the forethought to civilize them first—but not completely.

From the time I left the hospital until 1979, I led a normal professional life. Although my peripheral ties in the university and the discipline became attenuated, the faculty and students in my department rallied to my support, as did many close friends everywhere. As I have said, some people were unable to handle the new ambiguity in their relations with me, and I saw little more of them. Still others visited me in the beginning of my incapacity, but as illness stretched out into permanent disability, many ceased coming. To make matters worse, the dropouts felt guilty, and this made it even more difficult to reestablish ties. One friend rarely drops in on me at home, but he always visits me when I'm in the hospital—and he lives on my block. He is aware of this and is a bit puzzled by it himself. I suspect that he, and others, do this because the indeterminacy of my status ends when I go to the hospital, where I am no longer betwixt and between. There, I am categorically ill.

There was little ambiguity in my reintegration into the anthropology department, and the reception by some was so warm that I had to fight off secular sainthood. The seriously disabled face the curious dilemma that I described earlier. If they wish to relate to the able-bodied, they have to be stoical and uncomplaining; anything else drives the others away. But this often leads people to look on the disabled person as a hero, which is an embarrassment, imposing great strain and added ambivalence on social ties. I am not at all courageous—rather, I am a survivor, along with a good part of my generation. I have fought off canonization over the years because, despite its charms, I know that sacred people are treated with social distance, which would join forces with the impulse to avoid the disabled and isolate me further. Besides, I had devoted my entire professional style to preserving the mundane earthiness of my roots. In this way, I sought to avoid the fate of many iconoclasts: He who smashes idols risks becoming one.

My insistence on normality has been successful to some degree, and, in time, my office became the social center of the department because of my open-door policy. I literally have had to leave my office door open, because it is difficult to open it from a wheelchair and my voice isn't up to shouting "Come in" all day. But I also suspect that there was an agenda so neatly hidden that I have only recently discovered it. This is that I find it easier to handle encounters with small groups than with individuals. Individuals pose a problem of impression management, of negotiating the difficult process of normalizing the meeting, of negating the great contingency upon my identity. One-to-one encounters are also more problematic because the two parties may intrude upon closely guarded personal areas, something that usually doesn't happen in group sessions. These are settings in which my private sphere is safest, where I can flee into my professorial role and throw up barricades composed of a mixture of authority and humor. There is a closure in my openness, but I am not unique in this, for all social interaction is two-faced. Every juncture begets a separation.

Whatever the reasons, my office usually has visitors, and within my department I am very much in the middle of things. Fellow faculty members drop in to talk, sometimes on business but more often to swap complaints about life's unfairness to the middle-aged. The most frequent visitors, however, are the graduate students. They come during office hours and stay after airing their problems. My office hours tend to be additive, and, by the time they end, a small rump seminar may be in progress. This kind of easy and informal exchange has considerable educational value, but it also has its drawbacks. Many students understandably don't want to discuss their business in front of third parties, and I have had to make special appointments with them. In fact, it was their annoyance that made me aware of my own little social-distance game.

My relations with all students, as I have said, changed qualitatively because of my disability. But this has occurred also because I have aged to the point of being an academic grandparent. The parental generation in teaching consists of older assistant professors and younger tenured people. Although there are many exceptions, it is this group that imposes and enforces the rules. The older professoriat has seen students come and go, deans come and go, and rules come and go. They often have a wryer and more skeptical view of the academic enterprise than their younger colleagues, and they are long past the stage when one shows his or her high standards and scholarly rigor at the expense of students. The properly mellowed senior professor is one who handles the rules by exchanging a covert wink with his students. This, of course, is what grandparents do. They provide the system with just that measure of subversion necessary for its proper functioning. Age thus joined with physical impairment to bring me closer to the students.

My renewed immersion in teaching was part of this student-teacher pattern. It also fitted into the dominant characteristic of my first few years in the wheelchair, which was an attempt to cling to my old self through a denial of my disability. My denial was not so much a refusal to acknowledge what had happened, was happening, and will happen to me—there is no suppressing this—as an unwillingness to accept its consequences in my daily life. I knew damned well that I was in an abysmal state of health and did not especially try to hide it from others. What I would not admit is that I was socially or behaviorally disabled, and I worked twice as hard to prove that to others, and to myself. I have been fairly successful in my pose of normalcy, for people have been surprised to learn that my paralysis preys so heavily on my mind, as I rarely show it—but it does, and on the thoughts of all the deeply impaired. The act plays well at school, too, and one young colleague commented that she never thought of me as marginal, but rather as the ultimate Insider. She is right, for estrangement is situational, and this is another reason why my work is so essential to me.

Teaching is an important part of the anthropological life, but the profession's distinctive and defining feature is fieldwork, or ethnographic research. Often conducted in faraway places, among exotic people, and under difficult conditions, it lends anthropology a peculiarly romantic flavor. Yolanda and I had taken a full part in this tradition. We had worked among Amazonian tribesmen and African pastoral nomads, we had trekked through jungles, run rapids in wooden canoes, and eaten at campfires with tattooed Indians. We had seen things that few are privileged to witness, and that never will exist again; we had been to places that now are only memories; we had entered a universe that is on the verge of disappearing. But we would do those things no more.

By 1979, we were becoming increasingly restless and uneasy with our long absence from field research, a malaise that concurred happily with a growing interest in the social and cultural consequences of paralysis. My own changed circumstances first drew me to the subject, and my interest deepened as I read some of the literature. Most of it was written by sociologists, social psychologists, social workers, nurses, and physicians, and the quality of the work was uneven, ranging all the way from first-class social theory to some of the worst stuff I have ever read in the human sciences. And that includes some mighty bad things. It soon became apparent that no anthropologist had ever published anything on the social life of the paralytic, although they are a wonderfully researchable group—radically distinct from the general population and exotically different in their accommodation to social life.

I was not surprised at this disinterest in disability, for anthropology had its roots in the study of primitive—meaning nonliterate and socially and technologically simple—societies, an interest that still is the hallmark of the profession. The discipline wisely devoted its first fifty years of research to such groups, for they were disappearing rapidly and it was urgent that their ways of life be recorded. By the 1930s and 1940s, however, anthropologists turned increasingly to the study of peasant populations within nation-states, such as India, China, and Mexico. As the numbers of anthropologists increased after World War II, they branched out into different fields, and by the 1960s, many were doing fieldwork in American cities. Most of the urban research, however, was among minority groups of one kind or another, in keeping with the discipline's fascination with the culturally different. Anthropological attention also spread to the study of cultural beliefs and practices regarding illness. By 1980, anthropological research on the handicapped became firmly established with UCLA's work on developmental disability, although there were earlier isolated efforts, such as John Gwaltney's pioneering study of the blind in a Mexican village.[1] Yolanda and I decided that research on the motor-disabled suffered from a lack of anthropological methods and perspectives—an admittedly chauvinistic view—and we set out to correct the situation.

The first, and most essential, step in launching a research project is to find money, but it is necessary to learn enough about the subject to write a proposal. We sent in a prospectus for a small exploratory grant to the National Institute of Mental Health (NIMH), and, with its support, probed more deeply into published work and began our first interviews. Armed with our early findings, we reapplied to NIMH for a larger two-year grant and were surprised to find that the application had been referred to the National Institute of Neurological and Communicative Disorders and Stroke (NINCDS), another unit of the National Institutes of Health, of which we had never before heard.

The program, entitled "Social Relations and Microecology of Paraplegics," called for a two-year research period involving work conducted jointly by Yolanda and me, as well as by two graduate students working separately, under my supervision, on their own subprojects. Our purpose was to investigate and analyze the attitudes and behavior by which motor-disabled people adjusted to, or simply survived in, American culture, to explore the behavioral adjustments of paraplegics and their social circles to the fact of physical impairment. The program differed from most others in the field in that we would be studying people living in the community, either alone or in families, not hospital patients. It was distinguished also by an emphasis on in-depth interviews and what we anthropologists call the "participant-observer" technique of data collection. Using this method, we were not planning to conduct survey research, with its armament of questionnaires, but were going to learn through intensive and prolonged association with small groups of disabled people.

Participant observation is one of those ponderous terms that doesn't mean all that much; it refers merely to the anthropological practice of living amid the research population, taking part in their activities, watching what they do, and asking questions when one doesn't understand what's happening. This is less fancy methodology than simple necessity when studying an Indian tribe in the Amazon hinterland, where you can't pack up your notebook at 5 P.M. and go home. But it also has great methodological merit, for, unlike survey research, it allows the investigator to check statements of attitude and value against actual behavior. This is essential, as people often do not do what they *say* they should be doing, or even what they *think* they are doing. And it is further necessary because most anthropologists believe that one cannot properly understand cultural symbols outside their expression in social behavior, or behavior divorced from symbolic expression. *

In studying a society as complex as our own, it cannot be assumed that one person (or three dozen) speaks for all. In the

---

* I say this advisedly, for there is a strong movement among some anthropologists (and a horde of comparative literature scholars) that seeks to interpret cultural symbols in relation only to other symbols, and not to needs or acts. Inspired by a number of influential French writers, they analyze thought, cultures, and history as symbolic "texts." Under the appropriately awful labels of *deconstructionism* or *post-structuralism*, human culture's only purpose is seen as the construction of meaning. But they ignore the fact that, though we humans may have our heads in the clouds, the rest of us is embedded in dung. And before all else, cultural symbols mediate our actions as we go about the mundane business of work and procreation.

study of the position of the disabled in American society, it is necessary to recognize a whole series of contingencies. The adjustments of the affluent are of a very different order than those of the needy, and further variation occurs along lines of gender, age, race, ethnicity, and religion. In addition, the type of disability, degree of impairment, and age at onset also are important factors. It is clear that if one wanted to draw a broad profile of the motor-disabled, he would have to cope with dozens of separate categories. In order to get statistically meaningful results from each category, data on many individuals would be needed. Multiply their numbers by the number of categories and it is obvious that a sample of several hundred people is involved—too many respondents for close observation and intensive interviews. It's back to questionnaires and the shallow data of survey research. And this is what I refused to do. We tried instead to achieve a loose representativeness of sex and circumstance.

The participant-observer method was followed with a vengeance in my own case, for I was a full-fledged member of the group under study. For example, I spent five weeks on a rehabilitation floor during our research and learned far more about such facilities than I ever wanted to know. My two graduate student researchers found that the disabled were scattered all over the city, a problem in all urban research. This made it difficult to contact them and impossible to live among them in time-honored anthropological fashion. One of the researchers, Jessica Scheer, found a group of more than fifty wheelchair users—paraplegics and polio survivors—living on Roosevelt Island in New York City's East River, in a public housing development for middle-income people. It was as close to a community as one can find among the motor-handicapped, and she began research among them. She couldn't find an apartment in the complex, however, and instead became a daily visitor. Richard Mack was interested in finding black paraplegics, and Dr. Stanley Myers, director of the Spinal Cord Outpatient Clinic at the Columbia-Presbyterian Medical Center, allowed him to make contacts at the clinic with prospective informants. He then worked his way out from these people through their social networks to other respondents, and through them to disability groups.

We had earlier tried to establish a liaison at another hospital, an experience that was an eye-opener on some aspects of medical research. After much dickering, I was given the bottom line, which was that all publications derived from research at the hospital would have to include a staff member as coauthor. I gaped at the proposition and asked incredulously whether this was expected even when the staff member had taken no part in the research or the writing. I was assured that this was the custom in all hospital research, and suddenly I understood the impressive publication lists flaunted by medical people, as well as the fact that every paper seems to have six authors. This is a risky business, for a few very well-known people have had their reputations tarnished when the work of their subordinates was found to have been erroneous, and, in a few cases, faked. I refused, of course, to agree to such terms.

Fieldwork among American paraplegics and quadriplegics was a new venture for me in many ways. It was my first research on a complex society of any kind, and my first attempt to study my own culture. To complicate matters, I was investigating a population to which I belonged, a membership that I neither wanted nor accepted fully. There was a time when anthropology rested its claim for "objectivity" on its concern with other, often exotic, cultures. Since we did not share the values and biases of our subjects, it was argued, we saw them with a clear, nonpartisan eye. This was self-serving nonsense, for the truth is that our perceptions in ethnographic research are deeply affected by our personalities, by the language categories into which we sort out reality, by our education, by all the overburden of our own culture. And, as we get deeper into our research, the people's own interpretation of their culture provides an added coloration of our views, a further skewing of our perceptions. Our need to reduce all our data to a tidy system is just as much an attempt to cope with the sensory chaos of a world we do not fully understand as an exercise in science. And it is subject to the same errors and uncertainties.

Given the subjective element in all research by humans on humans, I was not too worried about my own objectivity. As I have said, my greatest risk is a temptation to believe that I know more than I really do; I have to be on guard continually against projection. At the same time, I possess a knowledge of paralysis that is "enjoyed" by few other researchers. My own disability gives me easy entry into the confidences of the handicapped, and it has been a useful guide when interviewing. At least I know what questions to ask. As for my values and sympathies, they are squarely on the side of the disabled, and I have been quite frank about this.

Although at first I resisted identification with the handicapped, my research drew me into their ranks. Yolanda and I took part in several disability organizations, and conducted interviews with paraplegics and quadriplegics in our suburban New Jersey area. It was a busy two years, filled with a constant round of interviews and meetings, conferences with the two student researchers, and my usual duties at Columbia. The grant freed me from half of my classes, but I still had a large number of graduate student advisees who had finished their research projects and were writing doctoral theses under my supervision. They included people who had worked among African pastoral nomads, aged Israelis, African agriculturalists, Latin American peasants, Amazonian Indians, Sumatrans, and a few other groups, and each student required lots of consultation. And as they completed chapters of their dissertations, I had to read and critique every line they wrote. By the time the project on disability ended in 1983, I was exhausted. I am still doing research on the paralyzed, but at a much slower pace, for things began to happen to me that punctured my personal myth that little had changed, that I could maintain the tempo and substance of my old life.

The degeneration of my central nervous system had been so slow that I was totally unprepared to cope with each new failure of function. Having decided to ignore my infirmities, I did little to educate myself on how to cope with disability, and I went to the doctor only when I had some specific complaint, which wasn't often. One of these problems was increasing spasticity in my trunk and legs, a change that aggravated my immobility. It was so bad at times that sudden spasms made me slip halfway out of the wheelchair, so I have had to use a seat belt for the past seven years. I have also been on an increasing regimen of muscle-relaxing drugs.

I did not take the time to consult my doctor in 1978 when I lost the ability to stand, and when it happened I had no idea how I would transfer from wheelchair to bed and back again. Fortunately, I was still receiving physical therapy, and my therapist got a sliding board for me. This is simply a smooth plank, beveled at both ends, that serves as a bridge from chair to bed or toilet. You get one end under your rump and slide along it. Simple, but I had never heard of such a board before. Why should I have? After all, I had until then been able to transfer by standing up and swiveling.

The end of my ability to stand up had a profound effect on my condition, both physical and social. First, reliance on the sliding board made me more dependent on others, for I wasn't able to handle it myself. Somebody always had to get the board under me and pull me along it. As a result, I lost a considerable amount of mobility. I no longer sat on a chair or sofa, but spent my entire day in the wheelchair. This, in turn, had a more serious repercussion, the development of sores on my backside; these are the bane of all who sit in wheelchairs all day, every day.

To the casual observer, it would seem that paralysis of the limbs is the principal affliction of the wheelchair-bound. This, however, is only the tip of the iceberg. As I have already said, the anal and bladder sphincters become inoperative, and a high spinal cord lesion affects one's ability to breathe. The autonomic nervous system controlling vascular processes is also affected. This means that the capillaries near the surface of the skin no longer respond to changes in temperature, which is another way of saying that one of the chief mechanisms for maintaining a steady body temperature is out of kilter. Moreover, the pores no longer open and shut or exude perspiration as they are supposed to do. All this reduces the body's ability to cope with temperature changes. As my paraplegia deepened and became transformed into quadriplegia, I suffered increasingly from summer heat. When the temperature hits eighty-five degrees, the air-conditioner has to be turned on. Winters are even worse, so much so that I cannot stay outside for more than a few minutes for fear of hypothermia.

Circulatory problems also underlie one of the major threats to the health of quadriplegics: skin breakdown. Decubitus ulcers, or bedsores, form when one sits or lies in one position too long. The constant pressure on one spot—usually one that is right over a bony prominence—prevents capillary circulation (already bad in quadriplegics) in the area, causing cells to die. The result is a small sore—no more than the size of a pinhead at the outset—which can expand into an ugly, gaping ulcer if the pressure is allowed to continue. Bedridden people often get sores on their heels, ankles, elbows, and shoulder blades, and the surest way to prevent this is by turning the patient every two hours or so. If that isn't practical, water mattresses help to inhibit the occurrence of the sores.

People in wheelchairs are prone to decubiti on the buttocks directly below the ischium, the bony protuberance in the backside, and under the coccyx, or base of the spine. Paraplegics can prevent them by pushing down on the wheels with their arms, thus lifting the buttocks off the chair, every twenty minutes or so. This allows the blood to recirculate into the compressed area, thus reviving the cells. But this technique is usable only by paraplegics or quadriplegics with good residual arm strength. Another way of fighting skin breakdown is by sitting on a good wheelchair cushion. I didn't know about any of these solutions, however, because I was determined that I would not let myself become a prey to my illness. But it was this very attitude that ultimately assured that I would.

Yolanda noticed the tiny sore on my left buttock shortly after it first appeared in 1980. Although we had not yet started our research, we knew a bit about bedsores, and she thought I should have it checked by the doctor, a suggestion that I dismissed immediately. Yolanda often urged me to go to a doctor but seldom went herself, so I shrugged off her warning as more alarmism. After all, it was only a little red spot; neither of us knew how dangerous these things really were. Besides, I was too busy to waste time in doctors' waiting rooms. I kept right on working, sitting in my wheelchair from 8 A.M. to 11 P.M. every day, using an ordinary cushion bought in a department store as my only concession to the sore. The ulcer thrived on such treatment, growing steadily in width and depth. Yolanda was becoming even more insistent that I seek advice, and my physical therapist joined forces with her.

During the winter of 1980–81, my general state of health began to decline markedly. I developed colitis, in part a byproduct of overwork, and a series of bladder infections. These combined with the infected ulcer to produce frequent low-grade fevers and left me vulnerable to a rather bad bout with flu in February. But I kept on working, driven now by an almost manic need for self-assertion and continuity. I was no longer so much denying the illness as defying it. Yolanda finally took matters into her own hands and told my doctor about the problem. He took one look and said, "Good God!"

I was admitted to the hospital immediately on an emergency basis for plastic surgery. The ulcer was already down to the ischium, and the infection would have entered the bone marrow if left untended much longer. The tiny sore that I had shrugged off had become life-threatening. To make matters worse, another small sore appeared on my right buttock. I had to wait a few days for surgery, however, because I had lost so much protein and blood through the sore that it took three pints of blood to get my red cell count up to normal. Three trips to the operating room were required to close up the two decubiti. On the first, the surgeon debrided the ulcer, a procedure in which he cut out the infected tissue in the lesion, which left me with a large cavity where a part of my buttock used to be. The surgeon also removed part of the ischium, which put the bone farther away from the surface, thus lessening the chance of a future break-down. On my second visit to the operating room a few days later, the surgeon took a flap of skin from the area adjoining the ulcer and grafted it onto the open wound. The third trip, a short one, was to clean out the ulcer on the right cheek and sew it up.

At the time of the surgery, my chest muscles had already weakened to the point where postanesthesia pneumonia could have been a real threat. My breathing had become more shallow, and I couldn't cough hard enough to expel fluids from my lungs or throat. And since I couldn't even blow my nose well, colds had become far more uncomfortable for me than for most people, a situation that has worsened since then. Pneumonia would require chest therapy to clear the lungs and perhaps even a respirator, and would be very dangerous. I talked over the problem with the surgeon, and we decided that I would undergo the surgery wide awake, in hopes that my tumor would provide an effective spinal block. He had an anesthesiologist standing by during each operation, however, to monitor my vital signs and intervene in case I went into shock. He would also be available to administer anesthesia if I felt pain; moreover, he was needed to administer a muscle relaxant to counter the increased spasticity that the operation surely would precipitate.

The first operation comes back into my memory in disembodied form, disarticulated from ordinary, conventional reality. It appears to me now as a fugue state, unconnected with any antecedents, and dreamlike in its outlines. This was due less to drugs than to the situation. I was lying on an operating table, chatting brightly with the anesthesiologist, while the surgeon was cutting out a bone-deep infected area, a procedure that should by rights have caused me excruciating, literally unbearable, pain. But I didn't feel even a twinge. I had not been nervous before the surgery, and the shot of Valium that the anesthesiologist put into my IV tube kept me relaxed and calm throughout it. Since I had nothing to do, I talked to the anesthesiologist, who wasn't very busy either. Like many physicians, he was an anthropology buff, although, unlike his cohorts, he didn't invoke general medical omniscience and pose as an expert. But he was well informed, for he asked me about Ruth Benedict's difficulties in winning tenure at Columbia some forty years earlier. I remember this question vividly, for I tried to answer while the surgeon was whacking away at my ischium with mallet and chisel, an episode made more bizarre by the sound of bone chips hitting the wall. It was an odd experience.

The three operations left me exhausted, but I was not nearly as stricken as I would have been if they had used a general anesthetic. Only once did they send me to the recovery room, a depressing environment when you are wide awake and can hear everybody coughing, spitting, and moaning. It reminded me of parties I had attended where I was the only sober person. On all three occasions, I had coffee and a snack after returning to my room, and also ate a light supper in the evening. The speed of my postoperative recuperation was wholly due to the fact that I had no anesthesia and, without it, no nausea or two-week hangover. The surgery mended fast and the graft took, but I had to wait out a six-week period in bed while it healed.

Although I spent all that time in a hospital bed, I had little rest. My next stop was in urological surgery for a prostectomy to facilitate the flow of urine; when the bladder cannot expel all of its contents, the stale urine becomes a breeding ground for infection. I underwent this procedure, too, without anesthesia. As soon as that was over, I was moved to the rehabilitation floor of the Neurological Institute, where I began physical therapy while still lying in bed. It was on this floor that the four four-bed rooms were made coed, or, to be more exact, were desexed. They were also terribly crowded, and there was room for only one visitor at a time to sit by a bed; the rest had to stand. It was a noxious place, the usual hospital woes magnified by the slumlike environment and further exacerbated by my imprisonment in bed. The only cheerful note came from one young man who had suffered brain damage in an auto accident, which left him paraplegic and unable to speak. But life wells up irrepressibly in the young, and he used to spell out indecent proposals to the nurses on a special board he carried. "What an awful thing to ask," they would say in feigned shock, as they hurried back to the nurses' station to regale the others. He would just grin.

During my last two weeks in the hospital, I was put back in the wheelchair for a few hours a day. After seven weeks in bed, I had to be sat up slowly, because if I rose suddenly, the blood would drain from my head and I would faint. They got me up by degrees by putting me on a tilting table and gradually increasing the angle of my body while monitoring my blood pressure. Once up, I was taken to the gymnasium twice a day for physiotherapy in an attempt to recover some of the strength I had lost during my long stay in bed. The recuperation took far longer than two additional weeks, however, and for the next three months I was still recovering from my decubitus ulcers, the plastic surgery, the prostectomy, and the weeks in bed. But my return to the wheel-chair was a release, an emancipation. All things are relative, and for friends who wonder how I can stand confinement to a wheel-chair, I ask, "Compared to what?" Compared to lying in bed, it's pure bliss.

There were some capabilities that I did not use during my nine weeks in the hospital, and consequently they became perma-nently weakened or lost. My arms had less strength and range of movement and my hands became weaker. I could barely push myself up from my seat to relieve pressure on my rump, a move made even more difficult by the fact that I was now sitting high on a Roho cushion. The Roho consists of sixty-four intercon-nected air cells that support the backside equally at all points, instead of letting all the body weight bear down on the area under the ischium. They are as good as one can find—they should be, at more than $250 each—but they are not infallible, for I got another decubitus three years later.

In January 1984 I came down with a bad case of flu, made more severe and debilitating by my neurological problems. I thus started the spring semester with my health at an ebb. It was not surprising, then, that about a month after classes began, Yolanda noticed a tiny telltale sore on my right buttock, the side where I had developed a small ulcer in 1981 that was simply stitched up. Of course, I should have gone to bed immediately and stayed there for twenty-four hours a day until the sore healed. But this could have taken weeks, for decubiti close up very slowly—people with large ones sometimes lie in bed for months—and I tried to put this off until the end of the semester. Needless to say, the decubitus got bigger and deeper, despite the fact that I spent hours in bed on most days. When the term finally came to an end in mid-May, the sore was already too large for slow healing. I entered the hospital in late May.

This ulcer had not reached the size of the first one, and a skin graft wasn't necessary. The surgeon only had to clean out the lesion, reduce the ischium, and close the wound. In order to promote the healing process and shorten the time I would have to spend in the hospital, the surgeon ordered a Clinitron bed. These beds, which in 1984 rented for $65 a day, have no mattresses. Instead, the patient lies floating on a bed of tiny ceramic beads that are suspended in air by a powerful blower system. All parts of the body are supported equally, relieving pressure on the bony prominences, thus allowing me to sit up in bed just a few hours after the surgery. One serious drawback of the bed, however, is that it generates a great deal of heat, and a bad heat wave arrived in June. Despite an old window air-conditioner, my small, one-bed room in the aging Harkness Pavilion became insufferably hot. To make things worse, the infection in the decubitus had entered my bloodstream, and I spent two weeks getting antibiotics through an IV in my hand. The infection gave me a low-grade fever, which joined forces with the heat of the room to render me very ill. I barely ate, and it was only during my last week in the hospital that I began to improve. Finally, four weeks after being admitted, I returned home, ready to start work on this book, for I was on sabbatical leave.

But things didn't turn out that way. Later that summer, to our total dismay, another ulcer formed on the surgical scar. I tried limited bed rest, but it got bigger. At the end of October, I went back for more surgery, this time staying for two months.

And so it was that I spent half of a six-month sabbatical leave lying in a hospital bed. When I returned home in December prior to confronting classes in January, I not only was weak from the surgery, but also totally enervated, my remaining muscle tone and strength sapped by the wonderfully comfortable—and seductively placental—Clinitron bed.

The bed eased all my muscle spasms, and my nights were spent in a state of relaxation that I had not known for years, and have not enjoyed since. During the days I sat up in bed, but at night the backrests were removed and my body sank into and floated upon a bed of air. I retreated into my shell, all outside sounds muted and overcome by the steady hum of the compressor pump and the soft murmuring of air. The bed lent itself to mental wandering and contemplation, and it was within it that I realized how this book should be written. It was the ideal ambience for remembering things past, for seeing my present circumstance as a revisitation of old and timeless experience. It was then that it occurred to me how much the bed was like being in a ship at sea, lulled and comforted by its gentle rolling and pitching, my senses dulled by the sound of air from the ventilators and the steady vibration of engines. The bed was like being inside a living body, an apt setting for thinking about basics.

To return to my narrative of 1981, I tried to be more careful after that first ulcer operation and took pains to lie down every afternoon for a half hour or so. The doctors' orders had been to sit for no more than two hours at a time, a regimen that would have made my further employment virtually impossible. Quite properly, I ignored their advice. Medical people have a penchant for looking primarily at the biological aspects of health, considering state of mind only when their diagnostic skills fail, as in the case of my "depression," which proved to be the spinal cord tumor. Their advice very often dooms the patient to social and psychological disability in the name of somatic "health," whatever that is. Accordingly, some medical people consider a paralytic to be doing well if he has no skin breakdowns, is not visibly depressed, and has clear bowels and bladder. Fortunately, the people at the Neurological Institute are not like that. They told me what the drill was, I snorted, and they shrugged.

Another capacity that I lost in the hospital and never regained fully was my ability to turn over in bed at night. During my first years in the wheelchair, I was able to do this easily, for my arms still were quite strong, and the wastage of my upper body was not far advanced. When my legs became more atrophied after 1978, I would bend my knees with my arms and push them off to one side, a position that would help me turn my body afterward. The increasing force of my spasms, however, made it ever more difficult to flex my legs, and I solved the problem temporarily by putting straps around the thighs, just above the knees, and pulling them over. Finally, even this gambit failed during the nine weeks of hospitalization, and I could no longer rotate in my old bed at home.

While I was in the hospital, somebody would turn me over every two hours during the night, something I could not and would not expect of Yolanda, who was already being pushed to the limit. The dilemma was resolved in two ways. We rented an electrically powered hospital bed, and a thin water mattress was placed on top of the regular one. For a couple of months, I could turn a bit at night by grabbing the bedrails and pushing and pulling my body from the left side to the right and back again. But this worked only if my legs and pelvis turned too, and my leg spasms soon made this impossible. I realized that from then on, I would have to lie in one position all night long, a grim prospect.

I had known for months that this night would come, just as I knew long in advance that there would come a day when I would never again be able to stand. Everybody rolls around during the night, whether half-awake or in deep sleep, for the body sends out discomfort signals when it remains too long in one position. I often wondered how I would be able to bear this sense of malaise and muscle tension when I could no longer move. If the night nurses missed a call in the hospital, I would awaken in acute discomfort after three or four motionless hours and ring for them. What would I do at home? Surprisingly, it was not as great a problem as I had feared. First of all, I soon broke out of the hospital-induced pattern of waking every few hours. Second, as the paralysis crept up my trunk, the discomfort messages became muted, then stilled. And third, the water mattress thus far has effectively prevented skin breakdown. I now sleep for seven or eight uninterrupted hours in exactly the same position. Of late, even my arms remain quiet.

As the tumor's pressure on the spinal cord in the lower cervical area has increased, my hands and fingers have become progessively weaker, stiffer, and less sensitive to touch. It is a process so gradual that it defies step-by-step description. The left hand weakened first, in keeping with the generally faster debilitation of my whole left side, but this was accelerated by my right-handedness and relative disuse of the left. I now have very limited tactile sense in the left hand, either for objects or for temperature. In 1981, I could still pick up light objects with my left hand, a capability lost today, and I still had opposability of thumb and forefinger, the great gift of human evolution. This is now gone, and my left fingers, stiff and barely movable, are curled inward toward the palm.

My right hand has followed the left by about a year or two, but it is now too weak to grip things tightly. It, too, has developed a tendency for the fingers first to curl in to form claws, then to close against the palm, but it is easier to prevent this in the right hand because it is used more. Nonetheless, I can no longer pick up a book, one of my few remaining forms of exercise. Rather, I pull the book to the edge of the desk and grab it between my two palms. I can still open it and turn the pages, although turning pages can be a slow and tedious affair. And when I can't do this any longer, I will do it with a stylus held between my teeth or use an automatic page turner. I don't worry about what I will do when the jaw muscles go, because people don't live long with that level of spinal cord lesion.

Despite the weakening of my hands and fingers, I still can write, after a fashion, and I still can type. After 1976, my handwriting became more slow and crabbed, but it remained legible.

All of my output of books and papers, from first draft onward, has been with a typewriter ever since the time I wrote my dissertation in 1954, and I have used handwriting only for signatures, short notes, and comments and corrections on student papers. In time, however, the writing became even slower and less decipherable. It also became fainter because of the weakening of the fingers, and I compensated for this first with soft-pointed pencils, then with ball-point pens, and, finally, with felt-tipped pens. I now write with a felt-tipped pen that has latex foam wrapped around the barrel, making it thick enough to grasp. But I do very little handwriting now. I can still do crosswords and double crostics—my only remaining addictions in a life once devoted to the pursuit of compulsions—and I can make a few cramped marginal comments on papers. But Yolanda writes all the checks and has my power of attorney, the ultimate symbol of trust in our pecuniary culture. For everything else, I type.

The ability to use a typewriter, and thus to ply my trade, also has changed. For the first five years that I was in the wheelchair, the electric typewriter served me well. If I had been a ten-finger touch typist, the infirmity would have been disastrous, but I was always a two-finger man, and my typing was unaffected. By 1981, however, the fingers of my left hand had become too weak and bent for typing. I then made a stylus out of a wooden pencil with latex foam wrapped around it. I held this in my left hand, hitting the keys with the eraser end. After about a year, my right middle finger weakened, too, and I had to make a stylus for my right hand. I then pounded away at the keyboard until my left hand became too weak to hold the stylus, calling forth a new technique. There is a simple little device called a "universal cuff," a strip of plastic that is wrapped around the palm and fastened by a Velcro strap. The plastic band is slotted, so that a fork or spoon handle can fit into it, allowing a person to eat without having to grip the utensil. I put a pencil into the slot and continued hitting the keys with the eraser end.

My typewriter days clearly were numbered, however, for my left hand soon became too rigid to hold the paper while my right rolled it in, and I was having great difficulty in adjusting the paper. I wrote letters with lines that marched uphill across the page. Moreover, my handwriting had become so unintelligible that even I could barely read it. As everybody who has ever written anything knows, this meant that I had difficulty making marginal and interlinear corrections, a necessary part of all writing. It was obvious that I could no longer hold back the new technology. It was time to buy a word processor.

Having earlier in this book held forth with enthusiasm about the potential of the computer for the disabled, it may seem curious that I waited until 1984 to buy one. The answer is simple: I became a convert only after using the machine. Prior to that, I regarded high technology as just one more blandishment of a maleficent economy bent on ensnaring, then impoverishing, its victims, and I saw computers as devices that ultimately would make us extensions of machines, rather than their masters. My troglodyte mindset was all part of a postwar reaction to my service as a radar, sonar, and radio technician, and my 1943 state-of-the-art training in electronics. Computers reminded me of the Navy.

After I bought the computer, my inexperience as well as glitches in the hookup convinced me initially that my misgivings had been accurate, but once I mastered the thing, I became dependent on it. People have bored me to distraction with effusions about their autistic romances with these machines, so I will mention only the value of their word-processing programs for a quadriplegic such as me. First of all, I can do all the typing, revision, and editing by hitting keys, without ever touching paper. And instead of producing the sloppy, cut-up, and scribbled-over manuscripts of the past, I now press PRINT, and out comes flawless copy. The only disadvantage of the machine is one that is common to all: It makes for prolix and discursive writing, ill-planned sentences, and spongy syntax. The ease of revision and editing encourages messy composition. Sometimes I tell people that I no longer write—I process words. These are, however, avoidable drawbacks, and a small price to pay for great rewards. The computer has been a liberation, an extension of my useful professional life.

For all practical purposes, my left hand is now dead, and the right is following slowly. The arms are weakening, too. Through a series of minute downhill steps, I came to the point where my arms no longer could propel the wheelchair. My speed and control suffered first, and one day I found that I was unable to push the chair over a three-eighths-inch ledge on the kitchen door sill. From then on, I had to be pushed in for meals. Several months later, I couldn't get over a quarter-inch lip on my bedroom door sill. The right wheel went over easily, but the left would go halfway up and then fall back. This kept me from getting from the dining room, where I worked, to my bedroom. I could have had little ramps made, but it was clear that they soon would have become useless, that even the slight upgrades would defeat me. So I was now confined during the day to the living and dining rooms, dependent on others to move me into other rooms.

Within months, I encountered new obstacles. In common with all old houses, our floors sag in places, creating little hills and valleys so small that they are undetectable to the normal person. But when you are using weakened arms to push a wheelchair across them, you get to know every one of them. I could have made a topographic map of our floors. Every month the hills grew higher, the valleys deeper, until one day I got stuck in one, the left arm too weak to get me out, leaving me able only to go around in circles. I had come to the point where I had lost the ability to move about, and I ordered an electric wheelchair.

I had put off buying a powered chair because pushing was good exercise, and I knew that as soon as I stopped, the loss of arm strength would speed up. As it was, I had waited too long, for I had completely lost mobility. If I needed to be moved even three feet, I had to call on Yolanda. This placed additional freight on our already overloaded relationship, and I often chose fixity to avoid becoming too much of a nuisance. The upshot was that I generally remained in the same place for long periods of time. Yolanda sometimes would go to work for several hours, leaving me in front of the word processor, where I stayed whether or not I wanted to stay. It's like being chained—literally—to one's desk. This is a good way to get work done, but my physical environs had become more constricted, the horizons of my body closer.

As the pressure of the tumor has increased in the upper thoracic and lower cervical regions, further weakening the chest muscles, my breathing has become increasingly shallow. At the present time, I have somewhat less than fifty percent of lung capacity, a figure that would be alarming but for the fact that people have far more capacity than they ordinarily use. Nonetheless, there are days of neurological ebb when breathing is labor, when I yearn to be able to take in a full lung of air, to feel that wonderful sense of revival that comes with a long, deep inhalation. I had never appreciated what a joy it was until I no longer could do it. Various portable respirators are available, and sometimes I am tempted to use one on the bad days, but this kind of dependency would only bring on more bad days—until I finally stopped breathing on my own altogether. I am not yet close to this point, but the shallow breathing probably is a major contributor to my chronic fatigue.

I already have mentioned my worries at the start of each academic year about my physical ability to deliver a lecture, and I faced that crisis again most recently on September 9, 1985. During the previous three months, my voice had become weak and reedy, lacking in resonance and timbre, and my breathing had worsened. This was in part the result of disuse during the summer, but, at a more fundamental level, it was due to the general lowering of my state of well-being during the two-operations year of 1984, a deterioration perpetuated in early 1985 by a series of bladder infections. These were corrected by surgery during the summer of 1985, and I had recuperated for two months, but still I had grave doubts that I would be able to carry my large undergraduate course on introductory anthropology.

My schedule in the fall 1985 semester included a graduate lecture course on kinship and marriage that met for two hours on Monday mornings, and the big undergraduate course, which was held each Monday and Wednesday afternoon for an hour and a quarter. I approached the morning class with some trepidation, for even though there were only twenty students, it

lasted two hours. I started out by encouraging them to partici-
pate in discussion, but this is hard to do on the first day, and I
wound up speaking for the whole time. After class, a series of
people came to see me during my one-hour lunch break, and by
one o'clock I felt totally talked out. I had some hope that my
undergraduate class would be smaller than in the past, for we
had added another section of the course. But more than a hundred students enrolled, only about twenty fewer than in 1983,
the last time I had taught it. To my surprise, the class went better
than ever. I was in good voice and rising spirits, and I had to
force myself to stop at 2:25 P.M. The following week, the ampli-
fier broke, and I even managed without one. The sense of mastery over my surroundings returned in full force. Sitting in my
chair, commanding the attention of the students, I no longer was
disabled.

My hospitalizations of 1984 delayed the start of this book
until the winter of 1985, but my fall 1984 sabbatical was not
entirely wasted. During the four months between hospital stays,
Yolanda and I wrote a new chapter for a second edition of
Women of the Forest,[2] our book on Mundurucu women. I also
did a second edition of my 1979 textbook on cultural anthropology, now retitled *Cultural and Social Anthropology: An
Overture*,[3] a job that involved considerable editing, cutting, re-
arranging, rewriting, and addition of new materials. The new
editions are now in print. As for this book, I began it in the
winter and spring of 1985, and did some of the main work on
the first draft that summer. It has been a feverish time, for I
work on the premise that I am running out of time. But these
same feelings of mortality have been goading me onward for the
past nine years, becoming by now more of a nagging presence
than a voice of doom. Paradoxically, my physical deficits have
helped me to get work done, as I am incapable of doing anything
else. I used to waste time in the pursuit of distractions, and once
painted our entire house in California to avoid writing a book.
All I can do now is read, write, and talk—which is what academics call "work."

After reading about recent advances in neurosurgery using

lasers, ultrasound, and more advanced microsurgery, I consulted in 1983 with a surgeon expert in these new areas, and he assured me that he could remove the tumor. At least this would keep me from getting worse. I became quite interested in the idea, although I knew better than to build up hope, and then I asked him how my lungs would handle long general anesthesia. No problem, he said. They would do a tracheotomy (place an air tube through the neck into the windpipe) and hook me up to a ventilator. Hopefully, they would try to get me back to breathing on my own after a few weeks. It was then that I saw there was a third possible outcome to the surgery. I didn't mind a choice between improvement and dying, but I had visions of it leaving me in a neurological twilight zone, like the last surgery. It was then that I saw that I had no remaining margin of error. I could not risk losing the capabilities left to me, and those things that made life worthwhile—Yolanda, family, work, and the constant pleasure of being a voyeur of the human scene. Besides, sixty was already old age for a male Murphy, and I thought that I would never live to amortize the operation. I turned down the suggestion.

As my body closes in upon me, so also does the world. My space is shrinking steadily, my mobility is lessened to a vegetal state. But my spatial placement seems to mean less to me now, for, no matter where I am, I am always anchored by inert flesh, caught within a faulty body, which is always sitting in the same chair, in the same position. Like all quadriplegics, I have a great fear of being left stranded and helpless, but my sense of self is otherwise shrunken to the confines of my head.

To fall quietly and slowly into total paralysis is much like either returning to the womb or dying slowly, which are one and the same thing. With all bodily stimuli to movement muted and almost forgotten, one gradually loses the volition for physical activity. This growing stillness of the body invades one's apprehension of the world. I have become a receptor in physical things, and I must continually fight the tendency for this growing passivity to overcome my thoughts. But there is a certain security and comfort in returning to my little cocoon every night, enswathed in a warm electric blanket, settled into a microenvironment consisting of one's essentials. It is a breach of communication with the toils of social ties and obligations, a retreat into a private cerebral world. And it is at these times that my mind wanders farthest afield. In such deep quietude, one indeed finds a perverse freedom.

8

LOVE AND DEPENDENCY

Enduring, diffuse solidarity, or love, in its most general sense in American culture is doing what is good for or right for the other person, without regard for its effect on the doer.
—David M. Schneider,
American Kinship:
A Cultural Account

Disablement is at one and the same time a condition of the body and an aspect of social identity—a process set in motion by somatic causes but given definition and meaning by society. It is preeminently a social state. And so it was that as my upper body functions became atrophied under the pressure of the growing spinal cord tumor, my social orbit shifted, my horizons shrank, my conduct of life became altered, and my sense of self underwent further deep transformation. The onset of quadriplegia, I discovered, had placed me in a new social dimension.

During my first two years or so in the wheelchair, I was reasonably self-sufficient. I needed help in putting on my pants, socks, and shoes, and I relied on Yolanda for transportation, but I could take care of most of my other mundane household needs without help. At Columbia, an assistant met me in the mornings, helped to get me from car to wheelchair, and took me up-stairs. I still was able to push myself around the building, and I got back and forth to the bathroom unassisted. But with the steady march of quadriplegia after 1981, this relative self-sufficiency ended radically, irrevocably, finally.

My inability to stand up made it hard to transfer from wheelchair to car. At first I used the sliding board, but this ma-neuver called for considerable effort by my helpers—meaning Yolanda, my assistant, or my friend and colleague Morton Fried, with whom I carpooled. Mort—a man of my generation and another survivor—and I had a curious kind of symbiosis in our travels. He suffered from recurrent vision problems, and on bad days I would warn him of various obstacles, such as stop lights and trucks. Our wives shuddered at this teamwork, but Mort and I thought it wonderful gallows humor, a neat varia-tion on the theme of the lame leading the blind. Mort fell ill in 1980, however, and Yolanda drove me herself, but managing the sliding board was a severe strain. We finally resolved the matter late in 1980, when we bought a van with an electric wheelchair lift. I am now lifted into the van, the chair is clamped down, and I am unloaded at our destination. In school, I require assistance for my every move.

It is hard for the average person even to understand, let alone empathize with, total physical helplessness. For example, in my third month in the wheelchair, I leaned too far forward one day, the chair tilted, and I slid out and onto the floor of our house. Falling is a constant dread of all chair users, for it robs one of mobility and access to help. Fortunately, I landed in a kneeling position in front of the phone, and I simply called the police to come and help me. On other occasions, my son, Bob, or the young men next door have picked me up and put me back in the chair. Help is not always close at hand, however, and I have gotten stuck in doorways or under furniture and have had to wait for an hour or two until somebody came home. This is less of a problem now that I have a powered wheelchair. But beyond these specific pitfalls, paralytics have a terror of being left help-

196

less and unable to communicate with others, especially if they live alone. A quadriplegic who does not have enough upper body strength to drag himself along the floor could lie for hours, even days, waiting for help. Among the security devices I use to fight these fears are an ever-handy telephone and intercom, my bulwarks against aloneness and immobility.

As the paralysis has crept upward and outward, numbing sensation and imprisoning movement, my dependence on Yolanda has become complete. A typical day starts when she awakens me at 8 A.M. and removes my night table, along with its array of support gadgetry—phone, intercom, TV remote, bed control, light switch, and water. She then washes the lower part of my body, a procedure that requires that she roll me first on my right side, then on the left. This is not an easy job, because my body is absolute dead weight, totally inert. After I am bathed, she rolls me back and forth again to get my pants on, following which she must roll me again to position a sling under my body. The sling is then attached to a Hoyer lift, a wheeled, hand-operated hydraulic crane that picks me up and moves me from place to place. Yolanda hoists me, then pushes the lift over to the wheelchair, and lowers me into it. Then I go into the bathroom for the next stage of my morning ablutions.

I brush my own teeth, using a toothbrush with a special thick handle, but Yolanda first must squeeze the toothpaste tube—my grip is no longer strong enough. Since I cannot lean forward over the sink on my own, she has to push my head over it so that I can rinse my mouth afterward. She then gets my shaving equipment ready and soaps the brush; I do the rest, using a razor with a built-up handle. There will come a time, not too far away, when I will be unable to shave myself, at which point I probably will grow a beard. After shaving, a tedious chore now, Yolanda bathes my upper body and washes my hair. Total time for the morning routine is about an hour, so I usually pass up the bathing and shampoo on the mornings when I go to school.

On the days spent at home, I use my power wheelchair, which enables me to get around the house, moving back and forth between computer and desk or anywhere else on the first floor. For school, however, I use the hand-propelled wheelchair, and Yolanda has to push me from the bathroom to the kitchen, where she serves my breakfast. We usually set out at about 9:30 A.M., driving across the George Washington Bridge, a short distance away, and on down to Columbia; the trip generally is only fifteen or twenty minutes. At Columbia, my assistant takes over and Yolanda leaves, sometimes to return home but often to go to work herself. Her part-time position on the faculty of Empire State College, a division of the State University of New York, takes her a few days a week to one of their facilities in either Rockland or Westchester County. After dropping me off, she often faces another forty-five minutes of driving.

Yolanda usually returns to Columbia to pick me up at about 4 P.M. Back home, I lie in bed for an hour or two, partly to relieve the pressure on my much-operated-on rump, partly to rest up after an often-frenetic day. From then until bedtime, I don't need much attention beyond being served dinner. I can still feed myself, using special bent-shafted forks and spoons with thick handles. My arm and hand movements when eating are awkward and labored, no doubt disconcerting or even repellent to some people when I eat out, but that's their problem, not mine. After dinner I usually read or watch television until bedtime, when Yolanda puts on my pajamas, hoists me from wheelchair to bed, sets up my night table, arranges my microenvironment, and settles me for the night. I very rarely call her on the intercom during the night, giving her eight hours of surcease from my care. It is for her a daily, weekly, monthly, yearly responsibility with no holidays or vacations, except for the periods when I go to the hospital. And then she feels obligated to visit me every day.

Beyond these burdens, Yolanda helps me in all my home undertakings. She gets books, changes computer disks, keeps my desk from passing from chronic messiness into total chaos, and does myriad little things that I no longer can do for myself. I don't need a constant attendant at home, for I can be left on my own for six hours or so at a time. For this reason, we seldom call on paid aides to look after me during the day, and Yolanda can go to her school, to New York, or shopping without worrying about me—though not for too long. Despite brief respites, she is tied down by me, her actions are severely limited by me, and my needs are never absent from her mind.

This has been but a gleaning of the main forms of my physical dependency on Yolanda, but there is a host of other little things. Just as an example, even though I still itch, I can't scratch myself in most places. I haven't been able to scratch my back or shoulders for the past ten years, and, as the range of rotation and extension of my arms has decreased, I no longer can scratch my scalp. For this I depend on Yolanda, and when she isn't around, I grit my teeth and wait for the itch to pass. My very ability to survive from day to day, to satisfy both major and minor needs, is in her hands. In a very real sense, we are both held in thrall by my condition—we are each other's captives.

Such dependency is much more than simple physical reliance upon others, for it begets a kind of lopsided social relationship that is all-encompassing, existential, and in some ways more crippling than the physical defect itself. It is not so much a state of body as a state of mind, a condition that warps all one's other social ties and further contaminates the identity of the dependent. Dependency invades and erodes the very compact upon which association between adults is premised. It joins forces with the damaged ego to subvert and corrupt social interaction. It assails and tests even some bonds that we hope are impermeable, such as those of marriage.

Independence, self-reliance, and personal autonomy are central values in American culture. One of our most persistent myths is that the country was built upon the efforts of singular individuals, men who had the daring and vision to found great enterprises, people whose fervent and unrelenting pursuit of personal gain or glory brought progress and prosperity to all. These men, so the myth goes, succeeded without help from government, or anybody else. They stood alone and—to use the current metaphor—they stood tall.

In our American myth, the lone individual brought order and justice as well as wealth. Several generations of Americans have been weaned on Western films that portrayed the solitary, history-less, and taciturn rider who arrives in a town dominated by the powerful and corrupt. In some films, there is a cabal of bad men, while in others there is a principal villain supported by a band of hirelings. In both types, the lone hero is pitted against a group, and groups usually are bad. In *High Noon*, the craven townsmen desert the marshal, leaving him to face insuperable odds. *Shane*, the best film in this entire genre, gives us a drifter of unknown place and past who takes the side of good farmers against bad cattlemen. Unassisted, he eliminates the opposition with his six-shooters, and then rides off into the sunset, repressing his love for Jean Arthur and his affection for her son. After *Shane*, other Westerns became redundant. The basic theme lives on, however, and in 1985, *Rambo* was a runaway box-office hit. Like Shane, Rambo is a man of few words, a reticence made more vivid by the fact that, when he does choose to speak, he is barely articulate. Rambo returns to Vietnam, where he had been part of a large American army that lost a war, and this time he wins a war all by himself. It is Shane gone bonkers.

The myth, of course, violates all the historical facts. Shane and the entrepreneur are a twentieth-century rewriting of nineteenth-century history, but what is most interesting about the *Rambo* movie is that it totally fictionalizes the Vietnam War, which ended only ten years before the release of the picture. It is probably for this reason that it appeals so strongly to those too young, or too dumb, to remember the conflict. Historical accuracy is of little relevance, however, for movies, myths, and dreams reflect a different reality, a substratum of desires and fears that haunt and shape our apprehension of the world. The hero is a natural figure who stands apart from group life and culture, while making the group whole again. He represents a dream of self-contained power, and of the ability to alter others while remaining unaffected by them. It is a myth rooted in physical power and validated by the hero's selflessness. He gives, but he receives nothing, not even love. The characters of Rambo and Shane are part of American values, projections of an uncertain and threatened masculinity, denials of emasculation, assertions of autonomy—all the very opposite of physical weakness and dependence, of disability. The disabled are indisputably the quintessential American anti-heroes.

Lack of autonomy and unreciprocated dependence on others bring debasement of status in American culture—and in many other cultures. Most societies socialize children to share and reciprocate, and also to become autonomous to some degree. Overdependency and nonreciprocity are considered childish traits, and adults who have them—even if it's not their fault— suffer a reduction in status. This is one reason why the severely disabled and the very old often are treated as children. The rules of reciprocity also affect the small-scale politics of social interaction, and we all know that in the subtle give-and-take of everyday life, those who accept largesse are under the power of the donors. The principles of reciprocity are built into all social relationships, and Lévi-Strauss tells us that they underlie marriage, the incest taboo, and the emergence of human culture.[1] These principles are also the mainstays of our personhood: To be a giver as well as a receiver is a hallmark of maturity.

It is for these reasons that escape from dependency has been a central goal of the disability political movement, and many handicapped people have discovered their own possibilities through going it on their own. For example, two young women whom we met during our research lived in a wheelchair-adapted apartment in a retirement housing project. One is a spinal cord– damaged quadriplegic with good upper body strength, although she has considerable atrophy of the hands. The other has cerebral palsy; she has moderate speech impairment and very limited arm and hand use. Both women use wheelchairs. Nevertheless, they both completed college, where they lived in dorms, and now were sharing an apartment. Each had a van, and the two did their own cooking and shopping, taking care of all their needs. The woman with cerebral palsy was unable to hold and use eating implements, so she was hand-fed by the other. Theirs was more than a viable living arrangement: It was a demonstration to the world that physical impairment need not diminish human dignity and integrity, and may even enhance it.

The problems of dependence versus independence, of contingency versus autonomy, are not restricted to American culture—they are a universal aspect of all social relationships. The ability to survive on one's own and to maximize self-determination are essential ingredients of the basic drive to live. We try to shape the social life around us, rather than to become its pawns or victims, and this involves the use of power, however subtle and gentle. The disabled, as I have said, have few such resources. Instead, they must seek social control by moral coercion, and social standing by the cultivation of admiration. But to become admired, one must be stoical and self-reliant. It's a hard act to maintain, and nowhere is it more difficult than in the disabled person's own family.

In the United States, and throughout the industrialized world, it is within the nuclear family that we have ties of greatest depth and wholeness. It is the forge and sustainer of our identities; it is where we find protection and reassurance; and it is where we exercise our most important social functions. But however much the family may be a haven in a heartless world, it is a refuge within which there often is more conflict and contradiction than in the realm we are escaping. Families, after all, are built on marriages, and marriages in contemporary society are built on quicksand.

Marriage is a truly peculiar institution, for it exacts a heavy toll upon those who enter into it. It provides them with sex partners, of course, but then the main reason why sex is in short supply is because of the restrictive rules of marriage. Its other benefits are equally questionable, for it has aspects of forced labor. The woman must be faithful to the man and bear and rear his children, and he must dedicate his sexuality and labor to the family enterprise. Beyond the need for such radical self-sacrifice, the marital tie can be abrasive, even corrosive; it is the arena of a constant battle to maintain one's identity, to preserve a delicate balance between fidelity to one's self and to one's spouse and children. In our society, good marriages find a middle ground in which a compromise is struck between mates, with neither becoming a tyrant or clone, but most unions never reach that ideal state. In the face of this power struggle, why do we persist in marrying? The answer, of course, is that the institution of marriage, which is found in every known human culture, serves society's purposes and not those of the individual. We humans serve society by work and breeding, and our cultures offer rewards to those who take part and unpleasantries to those who don't. These blandishments vary with cultural differences, but in our own mass society, single status invites disarticulation from social life and ineffable loneliness.

Power relations between husbands and wives in modern society are changeable and problematic. Gallons of ink and tons of paper have been devoted to this subject over the past twenty years, so I will touch upon only those aspects relevant to disability. The old sociological bromide that husbands/fathers are the leaders in practical decision-making in the family, while wives/mothers are the emotional centers, never was true in American society, for both spouses had in the past and continue to have important roles in each sector. Nonetheless, the husband's position as nominal head of the family was and still is seen as resting on his role as sole, or at least principal, breadwinner. As is well known, the general decline in American real family income during the past fifteen years has brought millions of women into the labor market, and the stereotypical couple of working husband and homebody wife is now a minority family form. As could be expected, when women added wage-earning to their contributions as housekeepers and rearers of children, their status in the family increased markedly; in cases where the husband lost his job and the wife provided the sole support, role reversals have taken place. Despite all these profound changes in

American society, an old sentiment persists: A man who stays home is a loafer and a failure, but a woman who stays home is a homemaker. Women *may* work, but men *must* work. And since a large percentage of the motor-disabled are not employed, they are economic dependents, supported by Social Security disability insurance and the incomes of their families. As would be expected, this dependency affects the social standing of men more deeply than women.

The presence of a severely handicapped, and consequently heavily dependent, member of the family has profound effects on its entire structure. The nature of that impact depends on the cohesiveness of the family prior to the disablement, but it also varies according to the sex and age of the impaired person. In his book *Passage Through Crisis*, published in 1963, Fred Davis reported on the sociological changes wrought upon sixteen families in which a child had contracted paralytic poliomyelitis.[2] Most of the families felt that they had experienced greater solidarity as a result of the illness, presumably because all the family members rallied to the support of the stricken child. The dependency of disability overlapped and reinforced the normal dependency of children, deepening the parent-child tie and strengthening the bonds among siblings. Many parents also reported that they had become more relaxed and lenient with all of their children, a reaction that they, themselves, attributed to their indulgence of the sick child and a wish to avoid favoritism. In sum, the relations of family reciprocity became stronger and the parental role mellowed, its disciplinary features softened.

An entirely different set of circumstances is set in motion when the disabled person is an adult. We have noticed in our research that, in several cases in which a disabled child grew to adulthood and stayed with his or her parents—a much more common pattern than among the able-bodied—the ties of childlike dependency continued. Loving support may have been necessary to the rearing of the handicapped child, but its persistence beyond childhood often results in emotional immaturity. The disabled young adult leads a sheltered existence in every way, and he or she often faces life with little experience of it, a deficit that is encumbered further by a built-in dependency orientation toward the world at large.

Much greater disturbances of the family system are set in motion when the husband or wife becomes disabled, for the usual authority and reciprocity roles are thrown completely out of kilter. One of the first casualties of paralytic impairment is often the couple's sex life, which may be terminated or altered drastically. Even when there is a successful adjustment, it takes time and more mutual understanding than exists in many marriages. My first encounter with the problems of paralysis and sex occurred in 1960, when I was teaching at Berkeley. When I reached the part of the course that dealt with the incest taboo (a subject then considered so tacky that only anthropologists would talk about it), a young man who used a portable respirator came up to me after class and said that he had run into a few cases of incest among people in the polio outpatient clinic at which he received treatment. All involved father-daughter incest in families in which the mother was stricken; the oldest daughter would take over the mother's household duties, which ultimately came to include sex with the father. The impact of the wife's disability on the family was so great that it eroded the incest taboo. In most cases, however, a disabled woman has a better chance of leading a customary sex life than a paralytic man, who may be rendered impotent by his condition.

These same sexual differences in the consequences of paralysis extend to other relationships. The identity of the male in the Euro-American world rests more heavily on work and occupation than does that of the female. There is no widely accepted social role of househusband, even for a disabled man. And in the reciprocal economy of the family, the handicapped man often loses his central function as main breadwinner. The loss of a man's power in sex and economics is echoed in all his other activities. The father finds that he must rely on moral coercion or reasoning when asserting paternal authority, for he has lost physical dominance. His functions in the household are diminished by his impairment in other ways. He no longer mows the lawn, fixes leaky faucets, makes bookshelves, paints rooms, plays catch with his children, walks the dog, or does any of the hundreds of things that once may have defined and symbolized his role as husband and father. He is around the house all day long, but he has no role or purpose within it; the home remains firmly in the domain of the wife, and she is off to work.

The lot of the disabled woman may be somewhat better, but it is in certain ways a bit worse. On the positive side, her identity and public standing depend less on occupation than does the man's, and it is more socially acceptable for her to remain at home. Moreover, if her disability is not too severe, she may continue to do light housework and cooking, and her role as the center of affection and guidance of the children may be little affected. There are, however, certain disadvantages in the woman's position. First, women are much more the victims of standards of beauty in this society, and they may be devalued more than are men for their physical blemishes. The other major penalty to females is that nurturance and personal body care of dependent family members is traditionally, and remains to this day, a wife's duty rather than a husband's; women are trained for this role, but men aren't. From this division of gender roles, a basic truth emerges—a disabled husband generally can expect more solicitous care from his wife than she could from him, if the roles were reversed.

It is on the level of such simple physical dependency that a disability becomes the dominant motif of a family's life, for all social relations become reorganized around it. The disabled adult sometimes has to be cared for like a child, depending on the severity of his or her condition. A quadriplegic with no use of hands or arms will require all the care that Yolanda gives me, and more. Husbands become part-time nurses, which goes against social conventions, and wives find themselves with an additional child, which doesn't. After all, lots of women have able-bodied, but highly dependent, husbands, and the ascendancy of the wife among retired elderly couples has been noted frequently.

Such shifts within the family structure often create strains so deep that it self-destructs. Quadriplegic husbands may rage against their own dependency and resent the consequent dominance of their wives. Sexual dissatisfaction is a common-enough cause of marital friction among the able-bodied, and it is endemic among the disabled. Failure to find alternate means of erotic expression may, however, be a reflection of a certain poverty and lack of variety in previous patterns of lovemaking. Spouses may also find themselves unwilling or unable to provide care for their mates, and may flee the marriage. I recall one particularly poignant story of a young wife whose husband supported her bravely through her early years with multiple sclerosis, but who later couldn't stand the physical and emotional strain of providing acute care. Racked with guilt, he divorced her and she entered a nursing home. He remarried but still visits her regularly.

It must be said again that feelings of guilt permeate, and sometimes bury, the families of the disabled. There is the basic, self-accusatory guilt of the impaired person, exacerbated by his or her guilt over being a burden. The able-bodied family members, for their part, feel guilty because they are intact and the other person is not. This may sound familiar, for all families are guilt networks; it is at once a source of both unity and dissension within them. But the guilts are far more intense in families visited by disability and, correspondingly, much more difficult to contain.

As refuges from cold anonymity, marriages bear a heavy burden in our fractionated, depersonalized society, and this is why half of American marriages founder. When they are additionally freighted with the problems of the handicapped, they frequently dissolve. As a matter of fact, the divorce rate is significantly higher among the disabled than in the population at large, especially among younger couples. But these are divorcing times. There was an era when the divorce of one of my friends would make me wonder, "What drove them apart?" Today I look at the long-married and ask, "What keeps them together?" The question is even more relevant when one of the partners is

207

physically impaired, and it is with this thought that I return to the history of my incapacity—and of our marriage.

I first met Yolanda in the physical anthropology laboratory at Columbia early in January 1950; she was there to study bones and I went there to study her. We went across Amsterdam Avenue to have coffee, and from that day we became inseparable. Yolanda was born in Warsaw of a Polish father and an American mother; when she was two, the parents separated and her mother brought her to New York City, where her mother worked first as a seamstress, then as a dress designer on Seventh Avenue. Yolanda's first language was Polish, but she grew up as an American girl, forgetting her Polish and her few traces of Polish culture. She never saw her father again and has no memory of him. After graduating from a Catholic high school of unfond memory, she attended Hunter College, partly at night, where she majored in anthropology. It was by this route that she reached graduate study at Columbia, the physical anthropology lab, and coffee with me.

Some twelve weeks later, on April 1, we were married. It was truly a wedding of April Fools, for my two teenage brothers were living with me after the death of our father and grandmother in 1948, and they continued to do so for the first two years of our marriage. Add to this the fact that I was a penniless graduate student, and it's understandable that her family was initially unenthusiastic about the union. Even our friends wondered how long it would last, for the first two years sometimes were stormy.

We managed to survive economically on my $110-a-month GI Bill stipend (worth a good $800 in today's dollars), supplemented by a bit of part-time work, and most of the strains originated in the usual difficulties of forging a union of the separate histories of two very different people. It is a tricky job under the best of circumstances, and life in a small Amsterdam Avenue walk-up apartment with two brothers was far from optimal. In 1952, however, we reached a turning point, the phase in an anthropology graduate student's career when one is expected to go off and do ethnographic field research. My brother David graduated from high school and won a scholarship to Columbia College, Peter went to live with my older brother John, and Yolanda and I went to the Brazilian Amazon to study the Mundurucu Indians. It was there that we got to know each other, or perhaps were able to create each other in our own minds—which is really the same thing.

It is impossible to convey the subjective experience of our year with the Mundurucu, for they are a people of a different age, living in a network of other meanings, seeing and hearing things that we did not comprehend. To be cast among them, as we were, marooned us in a world without form, sense, or predictability—until our research uncovered the hidden rationality of their actions. But this very effort subverted our own prior sense of reality, leaving us estranged from our urban American culture and with a highly incomplete grasp of that of the Mundurucu. This tension between separately constructed realities— and the shock of fully understanding the fact that much of reality is conventional artifice—is the essence of the anthropological experience and the forge of its altered perspectives.

Yolanda and I truly were babes in the wilderness, for our immersion in a radically alien culture stripped us of many of our preconceptions, even some of those about ourselves and each other. This left us to reconstruct and reconnect an altered universe of meaning in a colloquy in which we were the only participants. The sociologists Peter Berger and Hansfried Kellner have suggested that people maintain their views of reality through conversation; that is, they sustain, alter, and may even create, each other's sense of the real and the meaningful through mutual verbal assurance.[3] And nowhere, they write, is the conversation more intense, and extensive, than within the confines of a marriage. This was true with a vengeance in our case, for during the entire Mundurucu year, we rarely spoke English except to each other. And we spoke a lot. Seldom out of one another's presence for more than a few hours, we talked endlessly about our work, our respective pasts, our future, the lives and personalities of our new neighbors, and everything else. Our interpretation of Mundurucu culture was no doubt greatly influenced by the people's own views, but our reinterpretation of ourselves and our marriage was our own creation, unswayed and untested by other members of our own culture.

During our almost one year in Mundurucu villages, we rarely met non-Indians, and we had only one break, for the boat trip from Santarem on the Amazon River to the Mundurucu on the upper reaches of its tributary, the Tapajós, took us a full month. Isolated from "civilization," we became accustomed in time to the darkly tattooed visages of the people; to our thatch-roofed, bark-walled house; to the music of the sacred trumpets coming from the men's house; to the softly modulated tenor of social discourse. Our homeland was so distant that it had become irrelevant. We had mixed emotions about our situation. We missed the comforts of home and the variety of food and thought we had left behind, but we were drawn strongly to the relatively unriven quality of the culture around us. It was as if our step backward into a less segmented and differentiated social regime satisfied a secret atavism hidden in our inner selves.

In our isolation, we came to depend upon each other for emotional and physical support to an extent incomprehensible to most people. It was not an easy life. We slept in hammocks, our furniture consisted of packing crates, and we lived completely on the Indian diet. There were even moments of danger, such as the time we traveled a hundred miles through the rapids of the flood-swollen Tapajós River in a badly overloaded canoe, coming close to disaster a couple of times. These experiences— not unusual for anthropologists—forged an even greater degree of mutual dependence, which blended, in my case, with growing respect and admiration for Yolanda.

The year had its good moments. Yolanda remembers the warm sisterhood of the Mundurucu women, who often called upon her to accompany them to the stream. They would shake her hammock and say softly, almost musically, *"Iolantá, shet pin yun?"* ("Yolanda, are you sleeping?") And we still remember the hot afternoons when we would lie naked on the sandy bed of the stream and let its cool, clear waters run over our bodies while watching huge, white equatorial clouds pass over us. Or the pink and mauve dawns when we would sit behind our house drinking coffee, studying the tendrils of mist rising from the valley below—smoke, the people said, from the campfires of a mythic traveler. It was an enchanted land inhabited by living spirits and ancient memories, the landscape mapped and made sacred in the mythology of the people. Our year there was a kind of return to our own childhood innocence.

Our marriage, as it exists today, was born in the forests and savannahs of the headwaters of the Tapajós River, not in St. Paul's Chapel. It was an experience that could break a marriage, but it made ours. What emerged from the field trip was a tie based on deep and abiding mutual reciprocity and support. The great introversion and exclusiveness that has surrounded our relationship, the intensity of our "private sphere," arose from our months of isolation. Out of this came also a deep sense of loyalty, a bond that did not guarantee tranquillity but did produce durability under sometimes-grave strain. The union that developed was also a total one in that it covered all aspects of our lives; we constituted an entire social system for a full year, a little two-person outpost of American culture. But, perhaps most important, our long dialogue forged a common, shared view of what the world was like, who we were, and what we wanted out of life. We had become a unity.

The strength of this bond doubtless has been bought at the expense of an erosion of self, that strange metamorphosis by which the long-married become adjusted to each other, for each of us certainly has lost a large degree of spontaneity and autonomy. These adaptations, however, are no guarantee of domestic bliss, and some of our difficulties have arisen out of last-ditch efforts to preserve our residual identities, our personal private spheres. Georg Simmel was right. Total and intense marriages breed great expectations and deep intrusions, and ours often was disturbed by its seeming success. We have seen bad times, especially the extremely unhappy period when I sank into alcoholism, but even in the worst patches, we never have considered separation or divorce.

During the years of my illness, I have become transformed from active father and husband, the source of help and support for Yolanda and the children, to passive recipient of services. All the many chores I used to do around the house—from interior painting and appliance repairs to lawn mowing and leaf raking—have had to be done by somebody else. Bob took over most of the jobs, and Yolanda discovered new mechanical talents. My role has been reduced to that of technical adviser, although Yolanda and Bob probably think of it as kibitzer, a dispenser of gratuitous advice. Needless to say, I often think they are not doing things right—meaning my way—and I usually feel compelled to say so. This creates problems of face-saving when I'm wrong, but it makes for even greater difficulties when I'm right, especially if I have the stupidity to crow about it or let smugness show on my face. They insist on the right to do things their way, which should be every worker's privilege, but that would leave me with little to do. Yolanda often asks my advice, but Bob rarely does. He, too, has a masculine stake in craftsmanship, and he, too, cultivates managerial self-assertion, that male characteristic so irritating to modern women. As I have stepped back from household activity, I have felt at times as if I had been put on a shelf, sidelined. In light of this position of passivity and dependency, my role as the chief financial support of the family has acquired greater symbolic importance in my mind; it became a mainstay of my ego.

The sense of estrangement from family life followed in part from my loss of mobility. Words and acts that took place on the second floor, in the basement, or outside were beyond my range, and a conversation in the kitchen could be finished by the time I got there. Spontaneity and movement are assumed as axiomatic in the give-and-take of social life, so taken-for-granted that they remain unspoken, and family members had no idea that I felt left out. At one point I complained to Yolanda that Bob never said hello to me when he came home. What he did was enter the house through the kitchen, utter a general greeting, and go upstairs without passing through the living room—something he had been doing all his life but that I now felt as a slight. After Yolanda pointed this out to him, he took care to detour my way in deference to my sensitivities, now gone morbid. He has learned to be tactful toward me. In little ways such as this, my disability has affected the very style of family discourse.

It might be assumed that family members would become so adept in their approaches to the disabled person that all the ambivalence, uncertainty, and agonizing self-consciousness would disappear. While familial relations of the disabled seem superficially to be far smoother—less anxious and uncertain—than with new acquaintances or even friends, the problems merely have shifted to a different plane. Family ties, especially the marital bond, are emotionally more intense, more diffuse in content, and more encompassing than those with the world beyond. Each family is a kind of secret society, a closed universe of information and dependencies, and also of loves, hates, suspicions, and jealousies. Social relations within its precincts are charged with an ambivalence that commonly is magnified by physical handicap, and the hurts inflicted in the family are felt more deeply than any others. Distortion of interaction of the disabled with family members may be less frequent than with strangers, but it is far more pervasive and damaging, for it infects the very haven to which most people return for support, protection, and love.

Many strains have stolen into our marriage because of my disability, but they are occasional, not constant, and we get along better than most couples, whether with or without a disabled member. During the course of a day, I usually ask Yolanda for dozens of small services, over and above the main care she gives me. Since I know that she is overburdened, I generally hesitate to ask for things and feel slightly guilty about bothering her—a guilt that becomes added to that caused by my

213

damaged body. As a result, I am especially sensitive to the tone of her response. Do I detect a note of impatience? Is she annoyed? Is she overtired? Should I have asked her? Does that slight inflection say, "What in hell does he want from me now?" This is not completely a concoction of my imagination, for we have been married so long that we are thoroughly familiar with each other's rich subverbal vocabulary of tone, accent, stress, gesture, and facial expression. After all, we had learned in the Amazon how to communicate in part-sentences, half-words, and grunts. In my disabled mindset, however, I pick up the right cues but I alter and magnify them, interpreting a small note of fatigue as major resentment and reading rejection into a fleeting expression of annoyance. The anticipation of such responses, in turn, affects the way I phrase requests. There has crept into my voice at times an edge of querulousness that was never there before, and it bothers Yolanda almost as much as it does me. My voice often anticipates a possibly negative reaction, and, by so doing, sometimes begets it.

In certain ways and at certain times, we have become defensive with each other. Yolanda often reads a peremptory quality into my requests and answers by saying, "Can't it wait? I'm busy." My reaction is usually anger at her imputation to me of a demand for immediate compliance. She naturally responds in kind, which ordinarily would set the stage for a full marital tiff, were it not for the fact that we know each other, and the sociology of our petty disagreements, so well. There is usually a tacit truce, a moment of silence, and we resume conversation as if nothing had happened. Nonetheless, there is a heightened self-awareness and guardedness in our relations that wasn't there before, and that has reduced openness and spontaneity. Our very attempts to avoid conflict through increased tact and delicacy have become part of the problem, not its solution. To some extent, we are caught in the same hall of mirrors that warps my more casual contacts, but in a far more intimate way. And if this can trouble a relationship founded on thirty-six years of marital solidity, between two social scientists who have made a study of what they themselves are experiencing, then perhaps we can understand how devastating disability is to most couples.

Disability, dependence, and unequal reciprocity have eroded my leadership role in the family, and the life of the household now centers less on my strengths as a person and more on the weaknesses of my body. My waning authority was not, however, a sudden, precipitous drop triggered by illness, but another step in a gradual process that long predated my disability.

In the first two years of our marriage, dominance was important to me, for it was part of the assertive masculinity of young men. Many of the problems of our early period together stemmed from the kind of political struggle that goes on in all marriages, a gender war that often lasts for the life of a union, and sometimes shortens it. The acute phase of the struggle in our marriage came to a halt when we went to South America, for life there was reduced to its elements, mutual support became a brute necessity, and I discovered in Brazil that Yolanda is quite indominatable. Our battle of the sexes was never thereafter resumed on a grand scale, for I realized early that I probably would lose, especially if I thought I had won.

During the time when our children were young, we slipped into a marital pattern that was standard for the period: Yolanda stayed home and I worked. My job, however, did not remove me from the household as decisively as most employment does, since I did much of my academic work at home. I spent a lot of time with my family, which softened the division of labor by gender. Despite this closeness, my moral claim to family leadership had been undermined badly by my drinking; instead of being a source of protection and security to Yolanda, I became a point of vulnerability and uncertainty. Yolanda had become the stable center of the family and my crutch. After I stopped drinking in 1966, she at first had a hard time sorting out the altered family relations and the loss of her monopoly on righteousness. During one argument, she accused me of being "sanctimonious" and "holier than thou," a major change of pace. Abstinence can be almost as upsetting to marital tranquillity as booze.

When the children reached the ages of ten and eleven in 1968, Yolanda took a part-time teaching position in a high school. During that year, I had a sabbatical leave and stayed home to write a book, and she went to work. It wasn't complete role reversal, but I did prepare the children's lunch and performed many household chores. Yolanda's work did a great deal for her that had nothing to do with financial control, which we always had shared. Our bank accounts always have been joint, and I simply gave her my paycheck every month—the unconscious legacy of a tradition in which the women took the pay before the men could spend it in a bar. Work wrought a transformation in her self-image and a sharp upward reevaluation of her own abilities, for Yolanda found through working that she had great talents. She completed her master's degree in 1971 and taught at Fairleigh Dickinson and Seton Hall universities in New Jersey before taking her present position at Empire State College. And in 1974 we co-authored *Women of the Forest*, which was based on her thesis; she was launched into professional life.

In that book we said that there was more symbol than substance in Mundurucu male dominance, which was equally true of our own marriage. Yolanda gave me symbolic deference, always asking what I wanted for dinner, where I wanted to go, and so forth, but she never took a back seat in the big decisions. For example, when Charles Wagley, my old friend and former teacher at Columbia, called us in California in 1963 to ask whether I would be interested in returning, Yolanda answered the phone, heard the offer, and said, "Sure, we'll come back." She was right, but it put me in a bad bargaining position. And, like most husbands, I learned early that she usually had an unspoken preference; my job was to guess it or tease it out of her. This pattern has continued into my time of disability, but Yolanda's deference is now even more hollow, the range of choice ever more limited by my physical condition, my power to make decisions increasingly compromised by my inability to carry them out. In the subtle politics of the household, by a series of steps too small to be described, we have passed parity to a point where the balance of power rests with her. Our mutual dependencies are no longer equal, her increased authority is justified by her added responsibilities, and I accept all this as fitting and proper.

Yolanda finds it difficult to express her thoughts on the subject of my illness, for it forces to the surface a decade of repressed grief, inarticulate emotional conflict, and ambivalence toward me. Depression sometimes afflicts the families of the disabled more acutely than it does the impaired person. I reacted to the first news of my diagnosis with emotional numbness, but Yolanda went home and cried, something I am incapable of doing. As my condition has worsened, her bouts of depression have continued, breaking out in sudden urges to weep at awkward times and in unlikely places. This is not a clinical depression, a symptom of neurosis, but a perfectly sensible response to physical fatigue and a desperate situation. After all, she has a husband with an incurable, degenerative, crippling, and potentially fatal disease; it reads like a bad soap opera.

The inroads of physical fatigue combine with overcommitment and role conflict to produce anarchy in her everyday affairs. If there is a stable point around which the rest of her plans gravitate, it is my needs. She is able to adjust her work to my schedule because Empire State College stresses independent study and tutorial work rather than classes, so she sees students by appointment. The result, however, is a desperate scramble to fit in meetings at school, shopping, cooking, eating, homework, and caring for me—the kind of ratrace run by modern, thirty-year-old "super-moms." The adjustments, therefore, are tenuous and tentative, a house of cards in danger of falling apart at any time. Her plans are never clear and stable, and she borders on panic as she runs from one chore to another. In this way, disorder compounds fatigue, and she feels trapped in a maze without exits.

Under these circumstances, one might well ask why Yolanda doesn't quit her job. The answer is only partially economic. Inasmuch as we have thousands of dollars in medical expenses every year that are not covered by health insurance, the money isn't irrelevant. But, far more important, her career is her last remaining bastion, the only activity that is uniquely hers, the one involvement that defines her as a person in her own right rather than as an unpaid nursemaid. Her fifty-mile round trips to either Nanuet or White Plains, New York, are forms of escape from me and my troubles, a transfer into another world that has little to do with the stresses at home. And she loves her job. For all these reasons, we both feel that she must continue teaching, but she still fears that my needs eventually will force her out, leaving her a prisoner of the house—and me.

My dependency weighs upon her heavily, teasing out all the negative side of connubial ambivalence. She knows that it is not of my doing, but she resents it no less. And she feels guilty because she harbors these sentiments, thus compounding the problem. My own reliance on her goes far beyond the norm in marriage, and the care she gives me is less wifely than maternal. She does all the things that a mother does for her child, and we both have reacted against this mutation in our relationship by a measure of repressed antagonism toward each other. I want a wife, not a mother, I say—but she has also remained very much my wife. I really have a wife-mother now. But didn't Freud say something about this? Do I protest too much?

Some of Yolanda's annoyances have more elusive origins. She has a properly jaded view of the "courage" that so often is attributed to me, because she sees me at times when the veneer of cheer in the face of adversity disappears, and all the antagonisms and frustrations of the day spill out. She knows that I am putting on an act, but so also is she. Her public role is that of the cheerful wife, the self-sacrificing, uncomplaining helpmate of an invalid husband, although her backstage demeanor can be quite the opposite. In fact, I am grateful for her occasional angry outbursts, for the very thought of being married to a happy martyr is appalling; I feel guilty enough without that. In private, we are much more frank with each other, and, if this results in occasional spats, then this is a small price to pay for the honesty we always have tried to observe. Besides, most of our relating goes on without such friction.

Disability affects the entire family, but we have worked hard to avoid the kind of encystment and total closure that often ensues, for this infects the home with a pathology as bad as the impairment itself. Inasmuch as I have remained active and involved, the house has remained open to the children and their friends, and there has been little avoidance or awkwardness in my relations with the young visitors. Both children were devastated by my illness, but as they grew into maturity, Pam and Bob took separate paths. Yolanda and I feel acutely the fact that Pam needed our support most at a time when the extremities of my condition made us least able to give it. She left home early and married. Bob went to college and, after graduation, returned home, where he still lives. He has a managerial position in a New Jersey industrial plant, but is always available to help us. This is not, however, the reason why he remains single; rather, he is part of the new generation that defers marriage as long as possible. In time he too will leave, as all the younger generation must, and Yolanda and I will be left. But we will endure.

Throughout our ten-year siege, Yolanda has clung to me as I to her, for she has her own forms of dependency, and our need for each other has protected our bond in trying times. We have built around ourselves a universe; we have become extensions of each other; we have absorbed each other. But in some ways we have remained strangers, for in all our thirty-six years together, we have negotiated the tricky game of holding on to one another while not losing ourselves. There is always that residual inner self where we store private moods and memories, and where we dream strange things. And so Yolanda remains a mystery to me, but this is why the magic remains after these many years.

There is a special joy in old marriages. The intense passion of the early years is replaced by a no-less-erotic friendship, and the delights of exploration give way to the satisfactions of knowledge. If he has aged well, the man will have resolved his maturation problems enough to discover that women of mellow years often acquire far greater charm, an allure composed of a mixture of greater wisdom, increased self-confidence, and deepened sensuality. The long-married are fellow campaigners who have faced down an inimical world together; they are a cabal bound in common silence toward outsiders. They are people with a single history and a common reservoir of memories. They remember children born and grown to maturity, and they remember shared sorrow and happiness. Yolanda and I remember, too, our life among peoples who still suffuse their worlds with enchantment and wonder—and we remember, above all, each other.

As I drift more deeply into physical quietude, I look back at it all and would change nothing, for I have been overcome by a growing sense of inevitability, a feeling that paralysis has its own logic and meaning and that I am embedded irrevocably in this structure.

9

THERE'S NO CURE FOR LIFE

"We know how the Universe ends—" said the guide, "and Earth has nothing to do with it, except that it gets wiped out, too."

"How—how does the universe end?" said Billy.

"We blow it up, experimenting with new fuels for our flying saucers. A Tralfamadorian test pilot presses a starter button, and the whole Universe disappears." So it goes.

"If you know this," said Billy, "isn't there some way you can prevent it? Can't you keep the pilot from pressing the button?"

"He has always pressed it, and he always will. We always let him and we always will let him. The moment is structured that way."

—Kurt Vonnegut, Jr.,
Slaughterhouse-Five

It ain't over till it's over. —Yogi Berra

In my hospital reveries on illness and decline, I had a haunting sense of having rehearsed for the present in all my past years, of reliving my history in hyperbole, of undergoing a savage parody of life itself. I was caught in a process from which there was no escape, one that was so inevitable that I could not resist it, only watch, spellbound. In a perverse way, the progress of my physical degeneration seemed meet and proper, for in each moment of my existence were all my yesterdays and all my tomorrows. And my recapitulation of the past—and future—was not idio-

221

syncratic, my own private nightmare. Rather, it has been in some ways an enactment in exaggerated form of the course of all of social life.

Paralysis is an allegory of life and entropy, and my search for their relationship places me in the role of the shaman, who seeks to reconcile the sick person to his illness by placing it in the context of timeless myth and belief. And my narration bears an eerie resemblance to the myth-telling of the shamans of the Shipibo tribe of the Peruvian Amazon, who, reports anthropologist Daniel Levy, relate their myths while holding their bodies absolutely motionless. The quality of movement is conveyed by voice and style and comes alive in the imaginations of the audience. My task differs from the shaman's in that it seeks no cure, only comprehension. This, then, has been a quest for meaning in a world in which there are no absolute meanings of any kind. The disabled, for understandable reasons, are more interested than most people in the conundrums posed by their affliction. Some resign themselves to God's inscrutable design, or believe that they are being tested by Divinity for some special purpose. This attitude has the virtue of answering the query, "Why me?" and enhances the individual's importance in the cosmos. It also addresses the problem of human suffering that is the theme of the Book of Job, that ancient complainer who blamed God for his boils. Whatever may be the comforts of religion, however, science refutes the urge to put humans at the center of a purposive universe. All our evidence is to the contrary, for the only discernible purpose in the process of evolution of organic life is to fill every niche in the environment with appropriate protoplasm.

There may be no final meanings, but we do live in a meaningful universe; otherwise we could not abide intelligent existence. Our meanings, however, are humanly contrived and quite arbitrary, lending a spurious but useful legitimacy to our lives, our aspirations, and our actions. That they are of very human invention is amply shown by the data of anthropology, for cultures, in all their lavish variability, are frameworks within which the people of different societies make differing sense out of their existence, and endow it with value. Reality, as I have said, is in part a social product, a collective delusion that sustains and guides life, but it is a fragile illusion that must be revivified continually by social activity, and by conversation. Most meanings, then, make sense only in relation to the rest of the culture of which they are a part, and not to some set of transcultural attributes. This axiom is gospel to most social anthropologists.

Having said that the systems by which we make sense of the world and invest it with value are relativistic, let me back off a bit and add that some are more relativistic than others. The conservation and perpetuation of life are paramount values everywhere, although they are expressed in various ways. The taking of human life, at least within a society, is nowhere regarded lightly, not even in American cities, where it is a daily occurrence. Life is valued and guarded in all societies, and death, despite its inevitability, always threatens the social fabric.

Love—or at least some kind of deep, positive, diffuse, and enduring tie—is another universal. In all societies, whatever their social systems, the mother-child relationship is primordial, enduring, intense; it is the seed bed from which all our bonds with others sprout; it is a type of union that humans try to recreate in later life. The particular symbols by which different societies interpret love and death vary, but their underlying structures are much the same: Life and love stand opposed to death and alienation. They are, by virtue of their pervasive importance and universality, central themes in all societies.

Paralysis is a metaphor of this struggle, for it engages its victims directly in the battle against dissolution. The preceding chapters have been a litany of the breakdown of the brain's control over the body and, with it, the replacement of intelligently directed movement by mindless spasticity and vegetal inertness. Paralysis sets in motion a process of estrangement from others, from one's own body, and ultimately from one's self. It is a metaphor of death and a commentary on life. Disability does indeed have a meaning, albeit one that is bestowed humanly.

But it is a general meaning that goes beyond a particular ailment, and beyond one or another culture, to encompass social life as a whole. Disability concerns our irreducible humanity, and in the shifting sands of relativism, I believe that there are such things as Man and Woman. I do not subscribe to the anthropological assumption that all humans are alike mentally in a negative way, that we are blank slates upon which culture writes. Instead, I follow Lévi-Strauss's teaching that all members of the species *Homo sapiens* share a common mentality that is constituted positively, and Freud's lesson that there are constancies in the human condition, whatever the culture. What, then, are these universals?

To understand the meaning of paralysis for the human condition, it is necessary first to consider the nature of our species. Most thinking about human nature has suffered from a quixotic search for the least common denominator of the natural human, or, at least, of the natural, genetic, generic man or woman. The social-contract philosophers of the seventeenth and eighteenth centuries mused upon man in a state of nature, untouched by the benefits, or the maladies, of civilization. Rousseau thought these "natural" humans were essentially good, and Thomas Hobbes thought they were pretty awful; modern anthropological opinion would say that there is no such animal as natural man, and the question of whether people are good or bad is irrelevant, a matter of subjective judgment.

Most musing on the subject of human nature has taken the form of a search for inherited biological propensities toward certain kinds of behavior. This has often involved attempts to trace these presumed hereditary traits to lower orders of animal life, making the quest less a search for a distinctively human nature than an inquiry into our behavioral links to animals. Out of this has come the modern field of human sociobiology, and with it a resurgence of instinct theories of human nature. An examination of this school of thought would take us far afield, but I would suggest that its popularity stems from its simplicity.

To explain one or another pervasive human behavioral trait, you find similar traits among animals, and then assume a genetic tie between them. One doesn't need to locate the gene—that's another department. In this way, causality is hypothesized in cases in which there is only analogy. This produces sloppy science.

Most anthropologists reject this facile approach to an infinitely complex subject. It is not because they dismiss human biology as irrelevant, for we are most certainly a species of animal that has evolved from other animals. Rather, it is because our biological drives toward certain kinds of behavior are broad and diffuse, and are thoroughly molded and shaped by our socialization—that is, by culture. Anthropologists are also leery of theories that have tended to naturalize, and indirectly justify, some of humanity's worst features, such as war and homicide.

The most important aspect of human behavior is that it derives its organization and content in the interaction of our biological drives with culture. This is not a unilateral relation between culture and a passive, malleable psyche, as behaviorist psychology would have it, but a process of give-and-take between culture and very resistant, self-serving, and quite rational individuals. A "natural" human being, therefore, is a contradiction in terms. Moreover, since our nature lies around us as much as within us, it is more appropriate to speak of the "human condition" rather than human nature. This is consistent with the lessons of disability, through which we found that, in many ways, biology has less to teach us than has social anthropology.

Our exploration of disability has laid bare the great impulsion of impaired people to collapse into themselves, as opposed to their need to reach out to others. Nobody saw more clearly than Freud the generality among all humans of this struggle. Shorn of the elaborate vocabulary of contemporary psychoanalysis, Freud outlined a drama that is universal in human experience. It starts with the birth of the child, who is at first a self-contained and undivided entity, all of whose urges and strivings are directed to himself. The socialization of the child is a process of weaning him first from this primal narcissism through love for his mother, and then weaning him from the wish to possess the mother by her rejection of these claims. The child thus learns that he must practice self-denial and redirect his energies and longings to the larger society beyond the family. But this tearing away leaves scars. Through it, we humans find that all attachments beget severances, that our first love was denied to us, and that love has its negative side: It is profoundly ambivalent.

Much as we may repress from our conscious minds the memory of the Oedipal drama, it returns to us in masked form in dream, myth, and action. It comes back, too, in an urge to regress, to recapture our earlier pleasures and to regain the infant's whole and unriven universe. Every step that we take outward and upward goes against this gravitational force. But the allure is tempered by the apprehension that return to the mother is a surrender to her ambivalent side, a symbolic threat of re-engorgement and destruction of the ego. The pull of the past has a social aspect, too, for it offers an end to self-denial and the imperative of reciprocity. It is Lévi-Strauss's Promised Land, a dream "removing to an . . . unattainable past or future the joys, eternally denied to social man, of a world in which one might *keep to oneself*" [Lévi-Strauss's emphasis].[1]

The two faces of Eros, then, are outreach to others and a backreach, or inward journey, into our past—a surrender to the nostalgic lure of primal narcissism. Toward the end of his life, troubled by the inevitability of World War II, Freud completed his human cosmos by a final opposition: Life and Death. The drive for life found its ultimate counterpoint in a death instinct, an urge toward pain, darkness, destruction, violence, and war. But is there such a thing as an inborn death instinct? It is not one of Freud's more popular or persuasive ideas, although its most vocal advocate, Norman O. Brown, uses the concept in his book *Life Against Death* as an element in his apotheosis of the individual, his vision of a future world in which there will be no repression.[2] I prefer to interpret the death instinct as the return of Eros to its narcissistic origins, a return to the undifferentiated self, to an infantile existence in which the child is master of a universe that he indeed embodies. Humans have a primal urge to reestablish this "oceanic" feeling of infinitude and to dissolve the boundaries of identity. There is, then, no separate instinct for the oblivion of the self, for death. It is a part of a generalized life force that reaches out for total Being and finds Nothingness, for they are one and the same. This, I believe, is Freud's real lesson. It was also Hegel's.

The return of Eros to its hearth is accomplished at the cost of its outer connections. It is a turning away from others, a rollback of the outreach begun in the Oedipal struggle, an implosion of social relationships; death and aloneness are different facets of nonbeing. Just as there can be no irreducible human nature free of society's imprint—or scarification—so, too, can there be no human existence in total isolation. Alienation from others is thus a deprivation of social being, for it is within our bonds that the self is forged and maintained. This loss of self, however, is inherent in the social isolation of paralytics, who have furthermore become separated from their bodies by neural damage and from their former identities by society. Their plight is that they have become divided from others and riven within themselves.

There is, then, a constant process built into all our endeavors in which we must reach out, relate, and love, while taking care not to lose ourselves. It is paralleled by the counterpoint between the allure of regression and return to the mother on one hand, and the fear of passivity and loss of power that it entails on the other. In the final analysis, social life is made possible by keeping a delicate balance between falling inward and falling outward. The structure of all our moments is that we are constantly being pulled apart between the two. It is also the story of our lives.

This theme is brilliantly developed in Samuel Beckett's novel *Murphy*, a parable of the human condition and an epitaph for the Irish.[3] The book's protagonist is engaged in a search for the self, and for a prolongation and expansion of life. To accomplish this, he spends all of his days sitting naked in a rocking chair, rocking hard at first but gradually slowing down until all movement stops; with his body so negated, his mind can wander free, seeking oneness within himself. Murphy's lady friend, Celia, a streetwalker by profession, would like to move into marital respectability and is understandably anxious that he get up and about and doing:

> "I am what I do," said Celia.
> "No," said Murphy. "You do what you are, you do a fraction of what you are, you suffer a dreary ooze of your being into doing."[4]

And so it was that our hero tried to stay the leakage of life by naked immobility—a paralysis by choice—a technique later used by the soldier in *Catch-22* who sat still and naked in darkness as a way of slowing time, and thus prolonging his life, while escaping the carnage of war. The final double bind, Catch-23, is that all roads do indeed lead to Samarra, and total inertia and isolation are but other forms of annihilation of the self. This becomes manifest when Murphy takes a job as an attendant in a mental hospital and finds his own avatar in the inner solitude of the schizophrenic:

> The function of treatment was to bridge the gulf [between "reality" and "unreality"], translate the sufferer from his own pernicious little private dungheap to the glorious world of discrete particles. . . . All this was duly revolting to Murphy, whose experience as a physical and rational being obliged him to call sanctuary what the psychiatrists called exile.[5]

But catatonia is a guarded world—one that is almost as hard to enter as it is to escape—a reversed consciousness and separate reality that has its own bizarre initiation rites. In the end, death is Murphy's only redemption from the insanity known as culture. The message of this tragic comedy (or is it a comical trag-edy?) is contained in the pessimistic words of his mentor, Mr. Neary:

> . . . the syndrome known as life is too diffuse to admit of palliation. For every symptom that is eased, another is made worse.[6]

There's no cure for life.

Throughout my odyssey, I have said to myself that I have come this way before, for the journey has echoed life's rhythm and will have a familiar resonance to everybody. All of us dangle between loss of self and loss of others, between outreach to the world and regression into our interior being, between life and death. But the paralytic's sense of *déjà vu* is sharpened by a society that rejects him, devalues his person, and throws obsta-cles between himself and any chance for living within and realiz-ing its values. The paralytic's inertia is symbolic of death itself; he is life's negativity. He represents an inverse definition of wholeness; he is a living reminder of the frailty of the body. It is a powerful metaphor, and a very large number of disabled peo-ple have surrendered and live permanently under its deep shadow. This is a kind of premature death in life, but it is a realm also inhabited by legions of people whose only disability is that they have given up too soon.

The forces of life—of resurgent, hungry, searching, assertive Eros—are strong, however, and there are millions of people in various stages of physical decrepitude who have fought off the shackles of dependency and the gravity of despair to fight their way into societies that have suspended judgment on their very humanness. They reject the limitations imposed upon them and the constructions placed on their identities, and they are doing this through interrelating and work. They are carrying forward one of the great fights of our century for dignity and freedom.

My book *The Dialectics of Social Life*, first published in

1971, investigated the eternal contradiction between the norms and meanings of culture and the ebb and flow of social activity. It ended, as it had to, in a profane epithet of defiance against society and its tyranny of cultural forms. This book has continued the project, probing through an understanding of the ontology—the state of being—of the paralytic into the condition of the individual in society. And I found it to be encapsulated in its highest form in the battle of life's wounded against isolation, dependency, denigration, and entropy, and all other things that pull them backward out of life into their inner selves and ultimate negation. This struggle is the highest expression of the human rage for life, the ultimate purpose of our species— paralytics, and all the disabled, are actors in a Passion Play, mummers in search of Resurrection.

And so we can return to the question asked at the beginning of our exploration: Is death preferable to disablement? No, it is not, for this choice would deny the only meaning that we can attach to all life, whatever its limitations. The notion that one is better off dead than disabled is nothing less than the ultimate aspersion against the physically impaired, for it questions the value of their lives and their very right to exist. But exist we will, for if all other meanings and values are arbitrary and culturally relative, then the only transcendent value is life itself. Life is at once both its own means and its end, a gift that should neither be refused nor cast off, except in utmost extremity. Life is less a state than a process, a drama with an inevitable denouement, for quiescence and dissolution are the fate of everything. But the essence of the well-lived life is the defiance of negativity, inertia, and death. Life has a liturgy that must be continuously celebrated and renewed; it is a feast whose sacrament is consummated in the paralytic's breaking out from his prison of flesh and bone, and in his quest for autonomy.

The paralytic is, quite literally, a prisoner of the flesh, but most humans are convicts of sorts. We live within walls of our own making, staring out at life through bars thrown up by culture and annealed by our fears. This kind of thralldom to culture turned rigidified and fetishized is more onerous than my own somatic straightjacket, for it induces a mental paralysis, a stilling of thought. The captive mind misses the great opportunity given to us by the chaos of today's rapid social change. This is to free oneself of the restraints of culture, to stand somewhat aloof from our milieu, and to re-find a sense of what and where we are. It is in this way that the paralytic—and all of us—will find freedom within the contours of the mind and in the transports of the imagination.

NOTES

PROLOGUE: NIGHT SOUNDS

1. Claude Lévi-Strauss, *Tristes Tropiques* (New York: Atheneum, 1974), p. 317.

1. SIGNS AND SYMPTOMS

1. Talcott Parsons, "Definitions of Health and Illness in the Light of American Values and Social Structure," *Patients, Physicians and Health*, edited by E. G. Jaco (Glencoe, Ill.: Free Press, 1958); Talcott Parsons, *Social Structure and Personality* (New York: Free Press, 1964).
2. David M. Schneider, "The Social Dynamics of Physical Disability in Army Basic Training," *Psychiatry*, 1947.
3. Erving Goffman, *The Presentation of Self in Everyday Life* (New York: Doubleday, 1959).
4. Erving Goffman, *Asylums: Essays on the Social Situation of Mental Patients and Other Inmates* (Garden City, N.Y.: Doubleday-Anchor, 1961).

233

NOTES

5. Susan Sontag, *Illness as Metaphor* (New York: Farrar, Straus and Giroux, 1978).
6. Ibid., p. 68.

2. THE ROAD TO ENTROPY

1. Victor Turner, *The Ritual Process: Structure and Anti-Structure* (Ithaca, N.Y.: Cornell University Press, 1969).
2. Oliver Sacks, *A Leg to Stand On* (New York: Summit Books, 1984), p. 211.
3. William Ryan, *Blaming the Victim* (New York: Vintage Books, 1972).
4. See Sigmund Freud, "Mourning and Melancholia," *The Complete Psychological Works of Sigmund Freud*, edited by J. Strachey (London: Hogarth Press, 1957). See also Jerome Siller, "Psychological Situation of the Disabled with Spinal Cord Injuries," *Rehabilitation Literature*, 1969.
5. See E. A. Weinstein and R. L. Kahn, "The Syndrome of Anosognosis," *Archives of Neurology and Psychiatry*, 1950. See also Morton Nathanson, Philip S. Bergman, and Gustave G. Gordon, "Denial of Illness: Its Occurrence in One Hundred Consecutive Cases of Hemiplegia," *Archives of Neurology and Psychiatry*, 1952.

3. THE RETURN

1. Émile Durkheim, *Suicide* (New York: Free Press, 1966).
2. Arnold van Gennep, *The Rites of Passage* (Chicago: University of Chicago Press, 1960).
3. Oliver Sacks, *Awakenings* (New York: Vintage Books, 1976), pp. 231–32.
4. Robert F. Murphy, *The Dialectics of Social Life: Alarms and Excursions in Anthropological Theory* (New York: Basic Books, 1971).
5. Robert F. Murphy, *An Overture to Social Anthropology* (Englewood Cliffs, N.J.: Prentice-Hall, 1979).

4. THE DAMAGED SELF

1. Erving Goffman, *Stigma: Notes on the Management of Spoiled Identity* (Englewood Cliffs, N.J.: Prentice-Hall, 1963).
2. Constantina Safilios-Rothschild, *The Sociology and Social Psychol-*

ogy of Disability and Rehabilitation (New York: Random House, 1970).
3. Goffman, *Stigma.*
4. George Herbert Mead, *Mind, Self and Society* (Chicago: University of Chicago Press, 1934).
5. Maurice Merleau-Ponty, *The Phenomenology of Perception* (New York: Humanities Press, 1962).
6. Ibid., p. 81.
7. Gelya Frank, "Venus on Wheels: The Life History of a Congenital Amputee," Ph.D. dissertation, Department of Anthropology, University of California, Los Angeles, 1981.
8. Oliver Sacks, *The Man Who Mistook His Wife for a Hat*, pp. 42-52.
9. See Siller, "Psychological Situation of the Disabled."

5. ENCOUNTERS

1. Goffman, *Stigma.*
2. Horace Miner, "Body Ritual Among the Nacirema," *American Anthropologist* (1956), vol. 58, pp. 503–7.
3. John Gliedman and William Roth, *The Unexpected Minority: Handicapped Children in America*, edited by Thomas A. Stewart (New York: Harcourt Brace Jovanovich, 1979).
4. Finn Carling, *And Yet We Are Human* (London: Chatto & Windus, 1962).
5. Yoko Kojima, "Disabled Individuals in Japanese Society," *Rehabilitation World* (1977). For a cross-cultural survey of attitudes toward the disabled see Jane Hanks and L. M. Hanks, Jr., "The Physically Disabled in Certain Non-Occidental Societies," *Journal of Social Issues* (1948).
6. Erving Goffman, "On the Nature of Deference and Demeanor," *American Anthropologist* (1956).
7. Beatrice Wright, *Physical Disability: A Psychological Approach* (New York: Harper and Row, 1960).
8. Fred Davis, "Deviance Disavowal: The Management of Strained Interaction by the Visibly Handicapped," *Social Problems* (1961).
9. Georg Simmel, *The Sociology of Georg Simmel*, edited by Kurt Wolff (Glencoe, Ill.: Free Press, 1950), p. 312.
10. David Rabin with P. L. Rabin and R. Rabin, "Compounding the Ordeal of ALS: Isolation from My Fellow Physicians," *New England Journal of Medicine* (August 1, 1982).
11. See Betty E. Cogswell, "Self-Socialization: Readjustment of Paraplegics in the Community," *Journal of Rehabilitation* (1968).

12. Davis, "Deviance Disavowal."

13. Carling, *And Yet We Are Human,* p. 18.

14. Yolanda Murphy and Robert F. Murphy, *Women of the Forest* (New York: Columbia University Press, 1974).

15. Robert F. Murphy, "Man's Culture and Woman's Nature," *Annals of the New York Academy of Sciences* (1977).

16. See R. William English, "Correlates of Stigma Towards Physically Disabled Persons," *Rehabilitation Research and Practice Review* (1971).

17. M. A. Chesler, "Ethnocentrism and Attitudes Toward Disabled Persons," *Journal of Personality and Social Psychology* (1965).

18. Murphy and Murphy, *Women of the Forest.*

19. Victor Turner, *The Forest of Symbols: Aspects of Ndembu Ritual* (Ithaca, N.Y.: Cornell University Press, 1967).

20. Mary Douglas, *Purity and Danger: An Analysis of the Concepts of Pollution and Taboo* (London: Routledge & Kegan Paul, 1966).

21. Turner, *The Forest of Symbols,* p. 96.

22. Ibid., p. 99.

23. Ibid., p. 95.

24. Jessica Scheer, "'They Act Like It Was Contagious,'" *Social Aspects of Chronic Illness, Impairment and Disability,* edited by S. C. Hey, G. Kiger and J. Seidel (Salem, Ore.: Willamette University, 1984).

25. Richard Mack, unpublished manuscript, 1985.

6. THE STRUGGLE FOR AUTONOMY

1. Mack, unpublished manuscript.

2. Frank Bowe, *Rehabilitating America: Toward Independence for Disabled and Elderly People* (New York: Harper and Row, 1980).

3. Bowe, *Disabled Adults in America* (Washington: President's Committee on Employment of the Handicapped, 1983).

4. Ibid., p. 14.

5. Ibid., p. 4.

6. Ibid., p. 22.

7. Leon Festinger, *A Theory of Cognitive Dissonance* (Stanford, Cal.: Stanford University Press, 1962).

8. Bowe, *Rehabilitating America,* pp. 152–54.

NOTES

7. THE DEEPENING SILENCE

1. John Gwaltney, *The Thrice Shy: Cultural Accommodation to Blindness and Other Disasters in a Mexican Community* (New York: Columbia University Press, 1970).
2. Murphy and Murphy, *Women of the Forest*, 1986.
3. Robert F. Murphy, *Cultural and Social Anthropology: An Overture* (Englewood Cliffs, N.J.: Prentice-Hall, 1986).

8. LOVE AND DEPENDENCY

1. Claude Lévi-Strauss, *The Elementary Structures of Kinship* (Boston: Beacon Press, 1969).
2. Fred Davis, *Passage Through Crisis: Polio Victims and Their Families* (Indianapolis: Bobbs-Merrill, 1963).
3. Peter Berger and Hansfried Kellner, "Marriage and the Construction of Reality," *Recent Sociology* 2.

9. THERE'S NO CURE FOR LIFE

1. Lévi-Strauss, *Structures of Kinship*, p. 497.
2. Norman O. Brown, *Life Against Death: The Psychoanalytic Meaning of History* (New York: Vintage Books, 1959).
3. Samuel Beckett, *Murphy* (New York: Grove Press, 1957).
4. Ibid., p. 37.
5. Ibid., pp. 177–8.
6. Ibid., p. 57.

BIBLIOGRAPHY

Beckett, Samuel. *Murphy*. New York: Grove Press, 1957.

Berger, Peter, and Hansfried Kellner. "Marriage and the Construction of Reality." *Recent Sociology* 2 (1972), 50–72.

Bowe, Frank. *Rehabilitating America: Toward Independence for Disabled and Elderly People*. New York: Harper and Row, 1980.

———. *Disabled Adults in America*. Washington: President's Committee on Employment of the Handicapped, 1983.

Brown, Norman O. *Life Against Death: The Psychoanalytic Meaning of History*. New York: Vintage Books, 1959.

Carling, Finn. *And Yet We Are Human*. London: Chatto & Windus, 1962.

Chesler, M. A. "Ethnocentrism and Attitudes Toward Disabled Persons." *Journal of Personality and Social Psychology* (1965), 2:877–82.

Cogswell, Betty E. "Self-Socialization: Readjustment of Paraplegics in the Community." *Journal of Rehabilitation* (1968) 34:11–13, 35.

Davis, Fred. "Deviance Disavowal: The Management of Strained Interaction by the Visibly Handicapped." *Social Problems* (1961), 9:121–32.

————. *Passage Through Crisis: Polio Victims and Their Families.* Indianapolis: Bobbs-Merrill, 1963.

Douglas, Mary. *Purity and Danger: An Analysis of the Concepts of Pollution and Taboo.* London: Routledge & Kegan Paul, 1966

Durkheim, Émile. *Suicide.* New York: Free Press, 1966.

English, R. William. "Correlates of Stigma Towards Physically Disabled Persons." *Rehabilitation Research and Practice Review* (1971), 2:1–17.

Festinger, Leon. *A Theory of Cognitive Dissonance.* Stanford, Cal.: Stanford University Press, 1962.

Frank, Gelya. "Venus on Wheels: The Life History of a Congenital Amputee." Ph.D. dissertation, Department of Anthropology, University of California, Los Angeles, 1981.

————. "On Embodiment: A Case Study of Congenital Limb Deficiency in American Culture." Wenner-Gren Working Papers in Anthropology, 1984.

Freud, Sigmund. "Mourning and Melancholia" in *The Complete Psychological Works of Sigmund Freud.* Edited by J. Strachey. London: Hogarth Press, 1957. 14:243–58.

Gliedman, John, and William Roth. *The Unexpected Minority: Handicapped Children in America.* Edited by Thomas A. Stewart. New York: Harcourt Brace Jovanovich, 1979.

Goffman, Erving. *Asylums: Essays on the Social Situation of Mental Patients and other Inmates.* Garden City, N.Y.: Doubleday-Anchor, 1961.

————. "On the Nature of Deference and Demeanor." *American Anthropologist* (1956), 58:473–502.

————. *The Presentation of Self in Everyday Life.* New York: Doubleday, 1959.

————. *Stigma: Notes on the Management of Spoiled Identity.* Englewood Cliffs, N.J.: Prentice-Hall, 1963.

Gwaltney, John. *The Thrice Shy: Cultural Accommodation to Blindness and Other Disasters in a Mexican Community.* New York: Columbia University Press, 1970.

Hanks, Jane and L. M. Hanks, Jr. "The Physically Disabled in Certain Non-Occidental Societies." *Journal of Social Issues* (1948), 4:11–20.

Jackall, Robert. "Moral Mazes: Bureaucracy and Managerial Work." *Harvard Business Review* (September–October 1983), 118–30.

Kojima, Yoko. "Disabled Individuals in Japanese Society." *Rehabilitation World* (1977), 3:18–25.

Lévi-Strauss, Claude. *The Elementary Structures of Kinship.* Boston: Beacon Press, 1969.

240

————. *Tristes Tropiques*. New York: Atheneum, 1974.

Mack, Richard. Unpublished manuscript, 1985.

Mead, George Herbert. *Mind, Self and Society*. Chicago: University of Chicago Press, 1934.

Merleau-Ponty, Maurice. *The Phenomenology of Perception*. New York: Humanities Press, 1962.

Miner, Horace. "Body Ritual Among the Nacirema." *American Anthropologist* (1956), 58:503–7.

Murphy, Robert F. *The Dialectics of Social Life: Alarms and Excursions in Anthropological Theory*. New York: Basic Books, 1971.

————. "Man's Culture and Woman's Nature." *Annals of the New York Academy of Sciences* (1977), 293:15–24.

————. *An Overture to Social Anthropology*. Englewood Cliffs, N.J.: Prentice-Hall, 1979. Second edition: *Cultural and Social Anthropology: An Overture*, 1986.

Murphy, Robert F., Jessica Scheer, Yolanda Murphy, and Richard Mack. "Physical Disability and Social Liminality." (In press.) 1986.

Murphy, Yolanda, and Robert F. Murphy. *Women of the Forest*. New York: Columbia University Press, 1974. Second edition, 1986.

Nathanson, Morton, Philip S. Bergman, and Gustave G. Gordon. "Denial of Illness: Its Occurrence in One Hundred Consecutive Cases of Hemiplegia." *Archives of Neurology and Psychiatry* (1952), 68:380–87.

Parsons, Talcott. "Definitions of Health and Illness in the Light of American Values and Social Structure" in *Patients, Physicians and Health*. Edited by E. G. Jaco. Glencoe, Ill.: Free Press, 1958.

————. *Social Structure and Personality*. New York: Free Press, 1964.

Rabin, David, with P. L. Rabin and R. Rabin. "Compounding the Ordeal of ALS: Isolation from My Fellow Physicians." *New England Journal of Medicine* (August 29, 1982), 506–9.

Ryan, William. *Blaming the Victim*. New York: Vintage Books, 1976.

Sacks, Oliver. *Awakenings*. New York: Vintage Books, 1976.

————. *A Leg to Stand On*. New York: Summit Books, 1984.

————. *The Man Who Mistook His Wife for a Hat and Other Clinical Tales*. New York: Summit Books, 1985.

Safilios-Rothschild, Constantina. *The Sociology and Social Psychology of Disability and Rehabilitation*. New York: Random House, 1970.

Scheer, Jessica. "'They Act Like It Was Contagious'" in *Social Aspects of Chronic Illness, Impairment and Disability*. Edited by S. C. Hey, G. Kiger, J. Seidel. Salem, Ore.: Willamette University, 1984.

Schneider, David M. "The Social Dynamics of Physical Disability in Army Basic Training." *Psychiatry* (1947), 10:323–33.

Siller, Jerome. "Psychological Situation of the Disabled with Spinal Cord

Injuries." *Rehabilitation Literature* (1969), 30:290–96.

Simmel, Georg. *The Sociology of Georg Simmel.* Edited by Kurt Wolff. Glencoe, Ill.: Free Press, 1950.

Sontag, Susan. *Illness as Metaphor.* New York: Farrar, Straus and Giroux, 1978.

Turner, Victor. *The Forest of Symbols: Aspects of Ndembu Ritual.* Ithaca, N.Y.: Cornell University Press, 1967.

————. *The Ritual Process: Structure and Anti-Structure.* Ithaca, N.Y.: Cornell University Press, 1969.

van Gennep, Arnold. *The Rites of Passage.* Chicago: University of Chicago Press, 1960.

Weinstein, E. A., and R. L. Kahn. "The Syndrome of Anosognosis." *Archives of Neurology and Psychiatry* (1950), 64:772–91.

Wright, Beatrice. *Physical Disability: A Psychological Approach.* New York: Harper and Row, 1960.